COMPLETE

A-Z

Physics

HANDBOOK

COMPLETE
A-Z
Physics
HANDBOOK

Michael Chapple

3rd edition

Hodder & Stoughton

A MEMBER OF THE HODDER HEADLINE GROUP

A B C D E F G H I J K L M N O P Q R S T U V W X Y Z

British Library Cataloguing in Publication Data
A catalogue record for this title is available from The British Library

ISBN 0 340 87272 1

First published 1997
Second edition 2000
Third edition 2003

Impression number 10 9 8 7 6 5 4 3 2 1
Year 2007 2006 2005 2004 2003

Cover photograph: Stewart Cohen/Getty Images

Typeset by GreenGate Publishing Services, Tonbridge, Kent.
Printed and bound in Great Britain for Hodder and Stoughton Educational, a division of Hodder Headline plc, 338 Euston Road, London NW1 3BH, by The Bath Press Ltd.

CONTENTS

A B C D E F G H I J K L M N O P Q R S T U V W X Y Z

HOW TO USE THIS BOOK

The *A–Z Physics Handbook* is an alphabetical textbook designed for ease of use. Each entry begins with a one sentence definition. The entry then opens out with analysis, diagrams and examples. If relevant, the initial definition will be followed by a word equation, a definition of the unit and the symbols for the quantity and its unit.

Entries vary in length according to the amount of detail which needed. 'Magnet' needs little more than a clear statement whilst topics such as 'displacement–time graph' or the 'photoelectric effect' need an introduction, annotated diagrams, some development of the principles and worked examples. The intention has been to offer a focused account of the subject without leaving out any of the essential steps towards a new concept. The general brevity of the work encourages the reader into a habit of continuous revision while still working hard on less familiar matters.

Entries contain cross-references in italic to other entries; these either support or extend the ideas in the entry. These references encourage readers to deepen and extend their knowledge of the subject by tracing a pathway from an initial area of interest to something less familiar. A shorter book gives a clearer view of the whole subject than does a longer, heavier and more detailed text and this speeds up the difficult processes of learning and understanding.

One great advantage of the alphabetical order of the entries is that the book has no beginning. Just open it anywhere and start reading, then move on several letters forward or back or let the cross-references guide you. Occasional browsing can be a profitable and interesting enterprise.

The later stages of the revision process will be facilitated by the appendices which list the physical quantities whose definitions are frequently examined and the experiments which examination candidates are often asked to describe.

I hope that the *A–Z Physics Handbook* will be an invaluable study aid in the weeks and months ahead. Thank you for your time.

Michael Chapple

ACKNOWLEDGEMENTS

My first thank you must be directed to the many teachers and colleagues I have met and worked with during the past years who have collectively drawn my attention to the numerous detailed but significant points that are so relevant to a secure understanding of fundamental principles in physics.

Secondly, I should like to thank Ian Marcousé, the series editor, and the late Tim Gregson-Williams at Hodder Headline for their encouragement and advice during the writing of the book.

Thirdly, I should like to thank my wife for not noticing all those other things I haven't been doing.

Michael Chapple

absolute temperature scale: a theoretical temperature scale based on a theorem in thermodynamics concerning the efficiencies of heat engines. It is sometimes called the absolute thermodynamic temperature scale. It is identical to the ideal gas scale of temperature. The numerical values are fixed by setting the triple point of water at 273.16 K above the *absolute zero*.

Thermometers are calibrated against the *International Practical Temperature Scale*. This scale of temperature has legal force. It is defined by setting numerical values to several reproducible temperatures such as the triple point of water, the boiling point of water and the melting points of a number of pure metals and by defining how intermediate temperatures are to be measured.

Since both temperature scales use the same set of numbers and have the same unit (the *kelvin*), you may wonder if there is any difference between them. Imagine that you are able to work out a theory from basic principles for, say, the variation of the resistivity of copper with temperature. You will end up with a formula and the temperature T in the formula will be an absolute temperature. When you compare your theory with experiment, you will measure temperatures on the practical scale.

Temperatures may be measured from absolute zero (K) or from the freezing point of water (°C). Temperature differences are given in kelvins on either scale. (See *centigrade temperature scale*.)

absolute zero is the lowest temperature on the *absolute temperature scale*. It is the temperature at which it becomes impossible for a body to release energy to its surroundings. It is unachievable in practice.

absorption coefficient: see *attenuation*.

absorption spectrum: the dark absorption lines seen across a *continuous spectrum* after it has passed through a cool gas. The wavelengths corresponding to these lines are identical to the wavelengths in the *emission spectrum* for the same gas. The diagram on page 2 underlines the correspondence between the emission spectrum of hydrogen gas in a discharge tube and the absorption spectrum for cool hydrogen. The ' spectral lines' are images of the slit, each line has its particular wavelength. All the focusing paraphernalia has been left out of the diagram.

The spectrum of a star is an absorption spectrum. The continuous spectrum is the *total radiation curve* associated with the temperature of the hot surface of the star. The absorption spectrum, seen as a mass of dark lines, is composed of the various line spectra of the elements which are present in the cooler gases some distance above the star's surface.

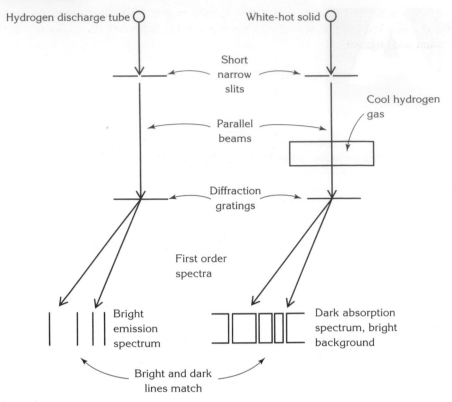

Hydrogen discharge tube   White-hot solid

Short narrow slits

Cool hydrogen gas

Parallel beams

Diffraction gratings

First order spectra

| | | | Bright emission spectrum

Dark absorption spectrum, bright background

Bright and dark lines match

Absorption spectrum

acceleration is rate of increase of *velocity*. It is a vector quantity represented by the symbol *a* and its SI unit is metre per second squared (m s^{-2}). The word equation definition is:

$$\text{acceleration} = \frac{\text{increase in velocity}}{\text{time interval}}$$

$$\text{m s}^{-2} = \frac{\text{m s}^{-1}}{\text{s}}$$

This word equation defines acceleration but not instantaneous acceleration or uniform acceleration. Instantaneous acceleration is the *limiting value* for the acceleration calculated over a very short time interval at the appropriate instant of time. It equals the gradient of the *velocity–time graph* at the same instant of time. Average acceleration only has meaning within the context of estimating, say, the average value of a force. It cannot be used in a precise calculation unless the acceleration is uniform, and then there would be no point.

Displacement, velocity and acceleration are related quantities. They are all vectors and all measured positively in the same direction. The three graphs on page 3 show how they relate to one another. A stone is thrown vertically upwards from the ground with an initial speed 29.4 m s^{-1}. It takes three seconds to rise just over 44.1 metres and another three seconds to return to the ground. Displacement is measured in the upwards direction from the ground; the first graph shows the changing values of the displacement during the six seconds. Velocity must be measured upwards too; the second graph shows the linear change with time from positive 29.4 m s^{-1} to negative 29.4 m s^{-1}. Velocity in the upwards direction decreases steadily in value. The upwards acceleration (measured in the same direction as displacement) is constant and negative.

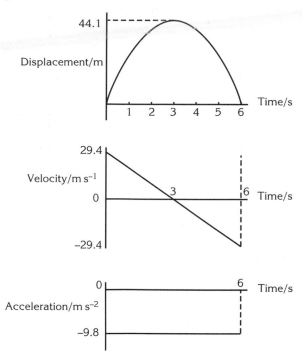

Acceleration is uniform if the velocity changes by equal amounts in equal times. These changes in velocity must be in the same direction and equal in magnitude. This can only happen if the body is moving along a straight line. The magnitude of a uniform acceleration is calculated from the following word equation:

$$\text{uniform acceleration} = \frac{\text{final velocity} - \text{initial velocity}}{\text{time taken}}$$

$$a = \frac{v - u}{t}$$

or, rearranged, $v = u + at$

The last equation is one of the *equations of motion*. It is an implicit definition of uniform acceleration.

An understanding of acceleration is essential in *straight line motion*, *circular motion* and *simple harmonic motion*. At A level, acceleration is uniform for straight line motion. For circular motion, acceleration is constant in magnitude but not in direction. For simple harmonic motion the magnitude of the acceleration changes constantly and the direction switches twice each cycle. Acceleration plays a key role in *Newton's laws of motion*.

Rate of change of *speed* is a scalar quantity which has no special name. It should not be called acceleration. Acceleration is a vector quantity, it is rate of change of velocity.

acceleration of free fall is the *acceleration* of a body towards the Earth's surface when it is acted on by one force only, its weight. This is the condition described, curiously, as 'weightlessness'. A satellite in orbit is acted on by one force only, its weight, and this weight is employed in providing the centripetal force.

If the Earth is treated, for the moment, as a perfect sphere of mass M and radius R, then the gravitational force F on a small mass m near the Earth's surface is given by

$$F = G.\frac{Mm}{R^2} = mg$$

where G is the universal gravitational constant and g is the gravitational field strength. (The Earth is not in fact a perfect sphere, being flattened near the poles, so the gravitational field strength g increases as we move from the equator towards the poles.) The unit for g defined by this equation is N kg^{-1}.

We can also write:

$$F = mg$$

where F is still the gravitational force on the mass m but where g is the acceleration of the mass under the action of this force. Here g is called the acceleration of free fall and its unit is m s^{-2}. But the Earth is spinning and the equation would be more correctly written:

$$F = \text{centripetal force} + mg$$

The *centripetal force* on an object of given mass is strongest at the equator.

The measured value for the acceleration of free fall increases from the equator to the poles for two reasons. First the radius of the Earth gets smaller so the gravitational field strength increases. Secondly the centripetal force on an object gets smaller towards the pole.

Acceleration of free fall does not differ much from 9.8 m s^{-2} the world over. Except near mountainous masses, its direction is towards the Earth's centre. If ϕ represents latitude and H the height in metres above sea level, then the value for g is given empirically by the equation:

$$g = 9.780(1 + 0.0053 \sin^2 \phi) - (3.086 \times 10^{-6} H)$$

For York, ϕ is 54°. Put $H = 0$ (which cannot be far wrong near a wide river) and we find $g = 9.814$ m s^{-2}. For London, ϕ is 51.5° and we get $g = 9.812$ m s^{-2}.

Experiments to measure the acceleration of free fall in the laboratory may be direct or indirect. A direct method involves using an object which is falling freely. An indirect method hinges on something like a *simple pendulum* or *helical spring* the period of which is a known function of g. Read the question; if you describe the wrong method you will generally be awarded no marks!

The direct measurement is a popular question and the time allowed usually about ten minutes. You would be expected to draw and label a diagram, list and describe all the relevant measurements and explain how the answer is worked out. The points that will be looked for in the diagram are the precise distance for which the fall time is measured and the circuitry for measuring fall time. Both elements should be clearly labelled. A skeleton circuit is usually sufficient for the time measurement.

Experiment: to measure the acceleration of free fall using an object in free fall

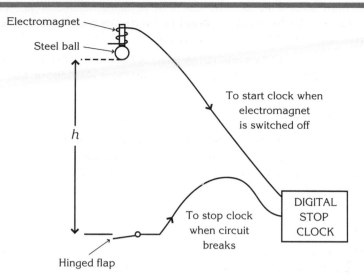

Apparatus

Set up the apparatus shown in the diagram. The distance h, the fall distance for the steel ball, must be labelled precisely. The digital clock is arranged to start when the electromagnet circuit supporting the steel ball is broken. It stops when the steel ball strikes the hinged flap at the bottom of its fall.

Measurements

Use a metre rule to measure the fall distance. Repeat this measurement twice and calculate the average value h. Measure the fall-time for one fall distance three times and calculate the average value t of these three measurements. Repeat this procedure for a range of heights between 50 cm and 2 metres.

Conclusion

Use the relationship $s = ut + \frac{1}{2}at^2$. Put $u = 0$, $s = h$ and $a = g$.
This gives us $h = \frac{1}{2}gt^2$. Draw a graph of $2h$ against t^2. It will be a straight line through the origin and the gradient will equal g.

Comments

Notice that repeating the measurements reduces random errors like some h values being slightly too large and others being slightly too small. But repetition does nothing for systematic errors like a clock that is too slow or a steel ball that sticks too long to the electromagnet. Systematic errors must be identified and minimised some other way.
You should add a word or two to explain how you minimise any parallax error when measuring h.

activity is the number of disintegrations per second which occur in a radioactive source. It is a measure of the strength of the source. The SI unit of activity, the *becquerel* (Bq), is defined as one disintegration per second. When new, school sources have activities in the region of 200 kBq. Some older sources have their activities marked in microcuries. The curie (equal to 3.7×10^{10} Bq) should no longer be used. (See *count rate*, *frequency*, *fundamental law of radioactive decay*.)

adiabatic changes occur when a system either gains or loses energy through working but not by heating or cooling. The usual reason for a change being adiabatic is that something happens too fast for any heat to flow. An example is the sudden expansion of air when a balloon pops. The *first law of thermodynamics* states that any increase in *internal energy* ΔU is the sum of heat ΔQ flowing into the system and work ΔW done on the system. When a balloon pops, the work ΔW done by the surroundings on the expanding air, is negative. ΔQ is zero. ΔU, like ΔW, must be negative. There is a loss of internal energy. The air mass cools down not because of heat transferred to the surroundings (driven by a temperature difference) but because of energy released when the expanding gas works on its surroundings. Similarly, a large mass of warm air rising from the ground expands adiabatically (because it is so big), it cools down, condensation follows and a cloud forms. It is sometimes convenient to resolve a change in the state of a gas into two component changes, one adiabatic, the other *isothermal*.

The pressure and volume of a fixed mass of gas vary during an adiabatic change in such a way that the expression pV^{γ} is constant, where p is gas pressure, V is gas volume and γ is the ratio of the specific heat capacities of the gas measured at constant pressure and at constant volume. Its value is 1.67 for a monatomic gas and 1.40 for a diatomic gas.

aeroplanes are supported in the air by an upward thrust on the wings generated by the different airflow rates over the upper and lower wing surfaces. The shape of the wing encourages the air speed v_U over the upper surface of the wing to be faster than the air speed v_L under the lower surface of the wing. *Bernouilli's principle* explains how this sets up a pressure difference and, therefore, a resultant thrust at right angles to the wing and upwards. This upward thrust is called lift.

The *free-body force diagram* here relates lift to the other forces acting on an aeroplane in horizontal flight at constant speed. The lift balances the weight. The frictional force of the air on the plane is called the drag. The drag balances the forward thrust from the propellers or jets. Archimedean upthrust is assumed to be small compared with the weight.

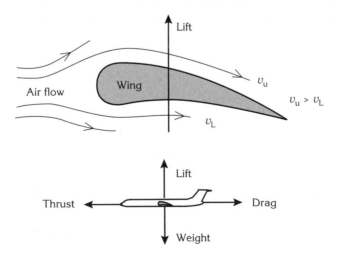

algebraic sum: the simple addition of two magnitudes. The sum of two scalar quantities is their algebraic sum, provided that the quantities in question **can** be added together. Mass (5 g + 8 g = 13 g) and electric charge (9 C + 7 C = 16 C) are added together algebraically. Electric charge, though a scalar quantity, can be either negative or positive.

Temperatures measured on the absolute scale can only be positive but temperature differences can be positive or negative. An equation like 300 K − 8 K = 292 K needs careful reading. The value 8 K is probably a temperature difference; 300 K and 292 K are probably temperatures. But we could be wrong and would need to know the context.

For some quantitites algebraic addition makes no sense, for example density or thermal conductivity. Vector quantities have directions as well as magnitudes and this makes their addition more complicated. (See *parallelogram law*.)

alloys are metals with new or enhanced properties formed by melting together and cooling two or more metals. The physical properties of an alloy depend on the metals chosen, the proportions combined and the process by which the alloy is formed. Brass is an alloy of copper and zinc.

alpha decay is the spontaneous emission of an alpha particle from a nucleus. An alpha particle is a stable particle formed from two protons and two neutrons. It is identical to the nucleus of the helium atom. The general equation for alpha particle decay is:

$$_Z^A X \rightarrow {}_{Z-2}^{A-4}Y + {}_2^4\alpha + \text{energy}$$

where X is the parent nuclide and Y is the daughter nuclide. The source of the energy released is the excess mass of X compared with [Y + α]. It is released in the form of kinetic energy which is shared by the two emergent particles in the inverse ratio of their masses. Since alpha decay is characteristic of heavy nuclei, the alpha particles carry off most of the energy. All alpha particles from a pure source have the same initial energy, characteristic of the source. See also *alpha radiation*, *radiation detectors*, *N–Z curve for stable nuclei*.

alpha particle: a stable particle formed from two protons and two neutrons. It is identical to the nucleus of the helium atom. Some radioactive materials of high atomic number emit alpha particles and the emergent flux of alpha particles is known as *alpha radiation*.

alpha particle scattering experiment: the classic experiment leading to the discovery of the nucleus and carried out by Geiger and Marsden under the direction of Rutherford. The essential features of the apparatus are shown in the diagram below. The alpha particle

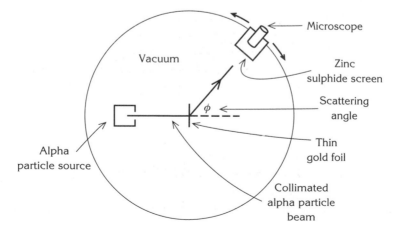

beam is narrow and parallel. The vacuum is essential; alpha particles travel no more than a few centimetres in air. The gold foil has to be very thin. The detector was a zinc sulphide screen, and scintillations were observed through a microscope.

The work involved three wholly distinct stages:

Stage 1 Having discovered that alpha particles pass freely through thin metal sheets, Geiger and Marsden checked the possibility of reflection. This was at Rutherford's suggestion. They found that a tiny fraction of the alpha particles were reflected.

Stage 2 This was the theory stage. Rutherford had shown a few years earlier that alpha particles are doubly-ionised helium atoms with enormous kinetic energy for their size. How could they pass through thin gold foil in their millions without producing any discernible effect in the foil. Equally, how could an atom of gold reflect an alpha particle without being torn from the foil?

Rutherford began working on the idea that the positive charge in an atom, and almost all of its mass, are concentrated within a small central nucleus. Electrons were somehow located near the edges so most of the atom would be empty space. An alpha particle which approached a nucleus head on would be slowed down to a stop and turned back by the coulomb repulsion, provided the nucleus of the gold atom was only about 10^{-14} m across. A larger nucleus would be broken up by the alpha particle. An alpha particle is identical to the nucleus of a helium atom and therefore very small. These ideas led Rutherford to the (simplified) scattering equation:

$$N_{\delta\phi} \propto \frac{1}{\sin^4 \frac{\phi}{2}}.\delta\phi$$

where $N_{\delta\phi}$ is the number of alpha particles scattered into the element of angular width $\delta\phi$ at the angle ϕ.

Stage 3 Geiger and Marsden checked this result by counting $N_{\delta\phi}$ for different angles ϕ, starting near zero and going up to almost 180°. They used different materials, different thicknesses and alpha particle of different energies. It was a painful few months. The outcome was the verification of Rutherford's nuclear model of the atom and a close estimate of the nuclear diameter.

This experiment should be seen also as an early example of a *scattering* experiment in which the nucleus is probed and its diameter measured. In the decades which followed, particle accelerators were built which have provided beams of electrons, protons and other sub-atomic particles to explore matter down to the details of proton structure. See *nuclear radius, scattering.*

alpha radiation is the outward flow of alpha particles from a radioactive source. Rutherford called it alpha radiation to underline the difference between the penetrating powers of this and the much more penetrating component, beta radiation. Gamma radiation was discovered some time later.

Rutherford used a spectroscopic technique to show that alpha particles are doubly ionised helium atoms or, as we would now say helium nuclei. They consist of two protons and two neutrons in a tight, stable bundle.

Alpha particles from a given source have roughly equal amounts of kinetic energy and therefore approximately equal ranges in air. Alpha particles from different sources have different energies and different ranges in air. Ranges in air are never more than a few centimetres. Alpha particles are stopped altogether by thin card or a thin sheet of aluminium.

Although alpha particles are highly energetic, their comparatively large mass (about 8000 times the electron mass) ensures a much lower speed which gives greater opportunity for interactions to take place with any atom in its path. Energy absorption is strong and penetrating power low. Although a strong burst of alpha radiation could cause skin burns, a more likely hazard is deep body damage to particular tissues if an alpha particle radiator were to be ingested.

Although their charge enables them to be deflected by electric or magnetic fields, the comparatively large mass and momentum of alpha particles ensure that any deflection is slight unless the field strength is high.

The most convenient way of detecting alpha radiation is with a *GM tube* attached to a scaler. The window of the GM tube must be made of thin mica, otherwise alpha particles cannot reach the detecting gases inside the tube. (See also *attenuation*.)

alternating current and voltage: the word 'alternating' implies frequent change of direction. In practice, the word is mostly used in connection with mains frequency currents and voltages. Although the term could include square (or any other shape) waves, it is normally assumed to imply a simple sine or cosine wave variation.

Typical equations describing an AC current and an AC voltage are:

$$I = I_0 \sin \omega t = I_0 \sin 2\pi ft$$

$$V = V_0 \sin \omega t = V_0 \sin 2\pi ft$$

$$\text{where} \quad \omega = 2\pi f \quad \text{and} \quad f = \frac{1}{T}$$

where T is the period of the signal in seconds, f is the *frequency* in Hz, ω is the angular frequency in radians per second, I and V are the instantaneous values of current and voltage respectively and I_0 and V_0 are their peak values.

ammeters measure electric current. Digital ammeters are accurate and cheap. There is no point in wondering how they work unless you have a special interest. They must be placed in series with the current to be measured, so that all the current flows through the ammeter and none bypasses it. The diagram below shows how a voltmeter and ammeter should be connected when measuring the current in, and the potential difference across, a resistor R.

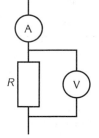

amorphous solids are either dust-like substances which lack any ordered structure at the atomic level, or substances like glasses and some polymers which are known to have no ordered structure at the atomic level. Some *polycrystalline* substances consist of such small crystals that they are effectively amorphous on a slightly larger-than-atomic scale. Amorphous means without form.

ampere: the SI unit of current. It is one of the seven SI *base units*. The symbol for the ampere is A.

The definition of the ampere is the end point of a sequence of ideas. One of the fundamental laws of electromagnetism, Ampère's law (which you do not need to know) is a rule for calculating the force between two short lengths of current-carrying wire. This rule is applied to the case of two infinitely long, straight and parallel conductors carrying currents I_1 and I_2 a distance d apart in vacuum. The force of attraction per unit length (for currents flowing in the same direction), F, is found to be as follows:

$$F = \left(\frac{\mu_0}{2\pi}\right)\frac{I_1 I_2}{d}$$

where μ_0 is the permeability of free space.

Now look at the SI definition of the ampere:

'The ampere is that constant current which, when flowing in two straight, parallel conductors of infinite length, of negligible circular cross-section, and placed one metre apart in vacuum, produces a force between the two wires of 2×10^{-7} newtons per metre of length.'

If these values are substituted into the above formula, we find that:

$$\mu_0 = 4\pi \times 10^{-7} \text{ H m}^{-1}$$

The ampere is defined then, by setting a particular and exact magnitude for the permeability of a vacuum, μ_0. (See also *permeability of free space, permittivity of free space, speed of light*.)

amplitude is the maximum displacement from its rest position of a body in *simple harmonic motion*. When the definition is extended to *wave motion* it is the maximum displacement of the medium when a wave flows through it. The definition can be carried beyond this to the maximum value of any oscillating quantity. The energy associated with either a simple harmonic oscillation or a wave form is proportional to the square of the amplitude. (See also *loudness*.)

amplitude modulation (AM) means linking an information wave to a carrier wave by simple addition of their amplitudes in preparation for transmission. The process was originally used when mounting a sound frequency wave on to a radio frequency carrier wave for radio transmission. See *modulation*.

analogue signals have *amplitudes* that can take any value between reasonable limits and that can vary with time. This changing amplitude with time carries information. A *sound wave* is an analogue signal. The term is used mostly to describe a voltage that varies continuously with time, recording the time variation pattern of some other physical quantity that the voltage signal represents. The output voltage of a thermocouple, for example, suitably amplified, will trace the temperature variations with time to which the thermocouple is exposed.

analogue-to-digital (A to D) conversion describes the action of a type of integrated circuit. The converter first samples the voltage levels of an analogue input signal at fixed (and very short) intervals of time. This timing is governed by a clock which drives two switches in antiphase. The sampling is usually achieved by charging a small capacitor to the instantaneous voltage level of the signal. This happens when the first switch is active. Next, the clock opens the first switch and closes the second switch. The voltage pulse from the charged capacitor is applied to a logic circuit whose output encodes the magnitude of the charge stored in the form of a binary number. An encoded binary number is illustrated in the diagram.

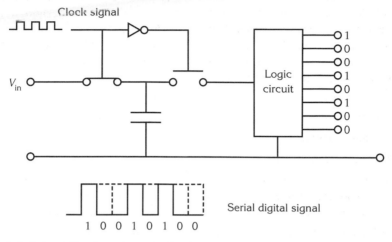

Clock signal

Logic circuit

V_{in}

1
0
0
1
0
1
0
0

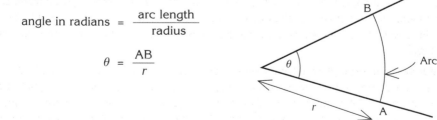

Serial digital signal

1 0 0 1 0 1 0 0

An 8 bit *digital signal* like that shown in the diagram can represent any integer of magnitude from 0 to 255 (i.e. 256 values), but not the fractional values in between. This means that the analogue signal has been 'quantised'. One consequence of this is that low voltages are represented with greater relative error than larger voltages: a digital signal of 233 units, say, is known to $\pm\frac{1}{2}$ unit, i.e. 1 part in about 250. But a digital signal of amplitude 3 units, also known to $\pm\frac{1}{2}$ unit, is known with an accuracy of about 1 part in 3. This disadvantage of the encoding process can be improved by *companding*.

The diagram shows the output of a digital-to-analogue converter as the parallel output from eight pins. The signals are usually converted to serial form for transmission. The two ends of the serial signal must be defined by coded start and finish pulses.

analogue transmission systems rely on the amplitude of the transmitted wave being a faithful copy in time of the amplitude of the original voltage signal. The advantage compared with digital transmission is simplicity; the disadvantage is the increased noise imposed by the transmission process, especially at the reception stage.

angle is an indefinable idea, it is a part of our language. But the magnitude of an angle can be defined, the unit depending upon the particular definition chosen. The commonly used unit, the degree, is defined as one complete rotation divided by 360. The diagram defines the magnitude of an angle in *radians*:

$$\text{angle in radians} = \frac{\text{arc length}}{\text{radius}}$$

$$\theta = \frac{AB}{r}$$

angular acceleration is the rate of increase of *angular velocity*. The symbol used is α. The SI unit is rad s^{-2}. At A level angular acceleration is constant, and in these circumstances the defining word equation is constructed by analogy with the definition of uniform acceleration:

$$\text{uniform angular acceleration} = \frac{\text{final angular velocity} - \text{initial angular velocity}}{\text{time interval}}$$

$$\alpha = \frac{\omega_2 - \omega_1}{t}$$

where ω_1 and ω_2 are the initial and final angular velocities respectively and t is the time interval for which the uniform angular acceleration operates. (See *angle, equations of rotational motion, radian, angular speed, pseudo-vector, torque*.)

angular displacement is the *angle* through which a body turns; it is usually measured in *radians*. The diagram shows a body before and after it has turned through an angle of 0.8 radians; the angular displacement is 0.8 radians. (See *equations of rotational motion, pseudo-vector, angular velocity*.)

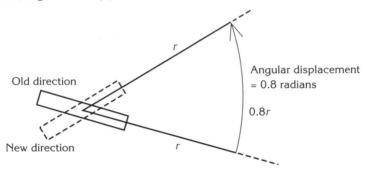

Old direction

New direction

Angular displacement
= 0.8 radians

0.8*r*

angular frequency is the *frequency* of a sinusoidal motion, whether mechanical, electrical or electromagnetic, measured in radians per second (not cycles per second). The symbol used is ω. Angular frequency in radians per second is related to frequency in cycles per second, f, by the relation $\omega = 2\pi f$.

angular momentum is the product of *moment of inertia* and *angular velocity*. It is a special kind of vector quantity known as a *pseudo-vector*. Most elementary particles have an intrinsic angular momentum and another component of angular momentum linked to nearby force fields. The latter component for electrons in atoms, explains the splitting of spectral lines when the source is placed in an intense magnetic field. Angular momentum is a conserved quantity.

angular speed is the rate of change of the angular direction of a radius joining a point on a rigid body to the axis about which it is rotating. The angular speed of a body moving along a circular path is the angular speed of the radius joining the body to the circle's centre. It is a scalar quantity. The SI unit for angular speed is radians per second, rad s^{-1}. The symbol used is ω.

Angular speed is defined by the word equation:

$$\text{angular speed} = \frac{\text{angle turned}}{\text{time interval}}$$

$$\omega = \frac{\theta}{t}$$

$$\text{unit of angular speed} = \frac{\text{radian}}{\text{second}}$$

Also

$$\text{angle turned in radians} = \frac{\text{arc length}}{\text{radius}}$$

$$\frac{\text{angle turned in radians}}{\text{time interval}} = \frac{\text{arc length}}{\text{time interval}} \times \frac{1}{\text{radius}}$$

$$\omega = \frac{v}{r}$$

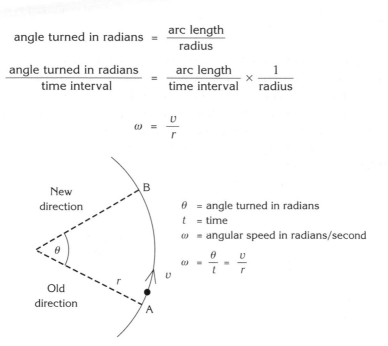

θ = angle turned in radians
t = time
ω = angular speed in radians/second

$$\omega = \frac{\theta}{t} = \frac{v}{r}$$

This last equation is an important relationship linking the angular speed ω for a body moving along a circular path of radius r, to the speed v of the body along the circular path. Notice that the quantity in the equation is speed, not velocity. Angular speeds should not be in degrees per second and certainly not in revs per anything. (See *angle, radian, angular velocity, pseudo-vector, angular acceleration, equations of rotational motion, centripetal force*.)

angular velocity is a vector quantity used in A level physics to define angular momentum and angular acceleration. Even then, the vector qualities of angular velocity and angular momentum are ignored. Several books use angular velocity when they mean angular speed.

annealing is the slow cooling of a body from a high temperature to room temperature. Tiny crystals, known as grains, grow during the cooling period. Slow cooling results in larger and fewer grains. The larger the grains within a metal, the easier and further it will bend without cracking. (See *quenching, tempering*.)

antimatter is a collective term for antiparticles. An antiparticle has mass in the ordinary sense – it will move in the direction of the force which pushes it. The special property of an antiparticle is that it will annihilate when it meets a corresponding particle. The combined rest mass energy of the particle–antiparticle pair is released in the form of two photons travelling in opposite directions. The quantum numbers allotted to an antiparticle (e.g. –1 for the lepton number of a positron and +1 for its charge) ensure that the conservation laws apply to any event in which an antiparticle is involved.

antinode: the point on a *stationary wave* where amplitude is a maximum. Antinodes form at those points where the component waves always arrive in phase. The formation of antinodes is an example of *constructive interference*.

Archimedes' principle states that a body that is wholly or partially immersed in a fluid is acted on by an upward force equal to the weight of fluid displaced. The upward force is called an *upthrust*. The resultant force on the body is the difference between the upthrust and the body's weight. This difference is zero for a floating body.

The first part of the diagram shows a stone suspended in air from a newton balance. The weight of the stone in air, W_{air}, is recorded. The second part of the diagram shows the same stone suspended from the same newton balance but immersed in water. The weight of the stone wholly immersed in water, W_{water}, is recorded.

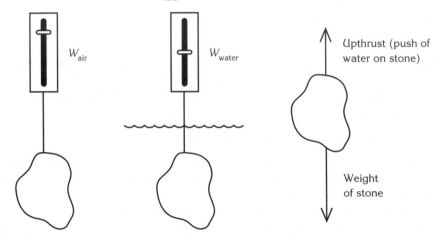

The water pushes up on the stone with a force equal to the weight of water whose place has been taken up by the stone. This is shown in the free-body force diagram.

Upthrust on stone $= W_{air} - W_{water} =$ Weight of water displaced

It is easy to calculate the masses of the stone and of the displaced water in kilograms from their weights in newtons. The volume of the displaced water can be calculated from its mass and the known density of water. This equals the volume of the stone.

Once both the mass of the stone and its volume are known, its density can be calculated.

The same stone could now be used to measure the density of any liquid other than water from a measurement of the upthrust on the stone when wholly immersed in the new liquid.

astronomy is the study of the universe as a whole. The main divisions are *astrophysics* (the study of stars) and *cosmology* (the study of galaxies and the structure of the universe). See also: *star, Hertzsprung–Russell diagram, end states of stars, galaxies* and *telescopes*.

astrophysics is about different kinds of *stars* and the varied processes that are going on inside them. The early life of a star is taken up with the slow gravitational collapse of an enormous cloud of hydrogen. The potential energy released ensures that the centre of the cloud gets hotter and hotter at a rate depending on the mass of hydrogen in the cloud. The mass of the Sun, 1.99×10^{30} kg, is known with some accuracy. The masses of about 200 stars (members of double star systems) have been measured directly and the rest of our knowledge is built on generalisations from this basic data. Even so, it is known that stellar masses range from about 10^{28} kg to about 10^{32} kg. Below this range, the temperature inside a star is never high enough for nuclear reactions to start. At best a star may release so much gravitational energy that, for a short while, it glows. These stars are known as brown dwarfs, but they have never been observed with any certainty. (The term 'brown' is curious – brown is not a spectral colour.) Above the high end of the mass range, the nuclear reactions would be too violent for any star to remain in a stable condition. Stars within the allowed range are called main sequence stars. See *Herzsprung–Russell diagram*.

Most of the energy released from main sequence stars derives from the so-called pp chain of reactions (where pp stands for proton–proton). This reaction chain is summarised in three equations:

$$\begin{aligned} {}^{1}_{1}p^{+} + {}^{1}_{1}p^{+} &\longrightarrow {}^{2}_{1}H^{+} + {}^{0}_{1}e^{+} + \nu_{e} \\ {}^{2}_{1}H^{+} + {}^{1}_{1}p^{+} &\longrightarrow {}^{3}_{2}He^{2+} + \gamma \\ {}^{3}_{2}He^{2+} + {}^{3}_{2}He^{2+} &\longrightarrow {}^{4}_{2}He^{2+} + {}^{1}_{1}p^{+} + {}^{1}_{1}p^{+} + \gamma \end{aligned}$$

This set of reactions can get started only if two protons approach one another with enough kinetic energy to overcome the Coulomb repulsion. This happens once the core temperature reaches about 10^{7} K. To increase the power output, two more conditions must hold:

1   the core temperature must be well above 10^{7} K and

2   there must be a high proton density in the core.

High mass stars have higher core temperatures and higher core densities than low mass stars: they 'burn' faster, they are brighter and their lifetimes are shorter. Hence, the mass of a star correlates with luminosity and, as we move up the main sequence to the brighter stars on the Hertzsprung–Russell diagram, stellar mass increases.

Most stellar energy comes from this burning of hydrogen. Once the hydrogen begins to become scarce, a new set of reactions involving the helium nuclei starts up. What happens next depends on the mass of the star.

A main sequence star can burn for a set life span. The later stages in the life of a star are described in *end states of stars*. The measurement of the distances of nearby stars is described under *parallax (stellar)*. The observed properties of stars are described under *star*.

atmospheric pressure usually means the *pressure* of the atmosphere at ground level. There are the large, the local and the microscopic points of view to consider. From the large point of view, air is seen as a compressible fluid with mass – a little more than 1 kg for each cubic metre. It is held on the Earth's surface by the Earth's gravity. Pressure decreases with height because the higher you go, the less atmosphere there is above to create the pressure to compress the atmosphere below. But, for a given temperature, pressure is proportional to density. So air density will diminish with height. It follows that:

$$\begin{bmatrix} \text{atmospheric pressure} \\ \text{at ground level} \end{bmatrix} \times \begin{bmatrix} \text{surface area} \\ \text{of the Earth} \end{bmatrix} = \begin{bmatrix} \text{weight of the} \\ \text{atmosphere} \end{bmatrix}$$

At a local level, air pressure varies a little from day to day but it is uniform enough at any given moment and over a large enough area not to have to pinpoint the place where it is measured. Its magnitude is found by balancing the thrust exerted by the air on an exposed surface against a spring of some sort or a column of mercury.

At the microscopic level, air is seen as a seething mass of countless molecules whose combined reflections on any surface they contact account for the thrust that measures the pressure. (See also *gauge pressure, mercury barometer.*)

atomic mass unit: an obsolete unit of mass equal to one sixteenth of the mass of an atom of the commonest isotope of oxygen. It has been replaced by the *unified atomic mass constant.*

atomic number, Z, is the number of protons in a nucleus. It is numerically equal to the number of extra-nuclear electrons in the non-ionised atom and indirectly governs the chemical properties of an element. The atomic number Z defines the element an atom belongs to. Isotopes of the same element have the same chemical properties except to the extent to which these properties depend on atomic mass. (See *N–Z curve for stable nuclei.*)

atomic theory of matter: the theory that matter is made up of particles, called atoms, instead of being infinitely divisible into ever smaller pieces. Although atomic theories have been discussed for more than two thousand years, the modern theory started with the discovery of the laws of chemical combination early in the nineteenth century. It is now universally accepted and the stress has been on understanding atomic and nuclear structure.

attenuation is the weakening of a beam of particles (or of electromagnetic radiation) by any material placed in its path. It may be due to scattering or to absorption processes. Either way, the degree of the attenuation is given by an attenuation coefficient.

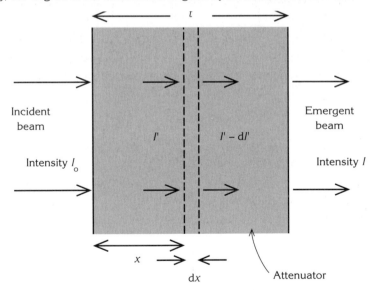

The diagram shows a beam of radiation or particles, incident at the surface of a sheet of material of thickness t. A thin section of thickness dx is shown a distance x from the incident surface. The incident intensity I_0 is the energy carried to the incident surface per square metre per second. I is the intensity of the emergent beam and I' is the intermediate intensity. Alternatively, I could represent the number of particles crossing unit area per unit time.

If dx is small, then:

$$dI' = -\mu I' dx$$

where μ, defined by this equation, is the linear attenuation coefficient for the material exposed to this particular beam. The equation can be integrated over the full thickness of the specimen to give an exponential law:

$$I = I_0 e^{-\mu t}$$

It is often more convenient to measure the mass per unit area of the material than to measure its thickness, particularly when very thin. In this case the linear attenuation coefficient μ is replaced by a second quantity μ_m, called the mass attenuation coefficient.

X-rays are often identified by the working voltage of the tube from which they come. For X-rays up to about 100 kV, attenuation is primarily a consequence of the *photoelectric effect*. The efficiency with which a material absorbs X-rays by this mechanism increases in proportion to Z^3 where Z is the atomic number. This is why bones show up so well on X-ray photographs. At around a million volts, Compton scattering takes over and this mechanism is independent of atomic number. This is why mega-voltage therapy is used to destroy offending tissue.

For *alpha radiation*, absorption in solids is too immediate for the *exponential* law to be of much interest. It works well for beta and gamma radiation. There is the complicating problem with *beta radiation* that the electrons produce X-rays when suddenly stopped in hard matter. These X-rays, known as bremsstrahlung, cause a severe deviation from the exponential decay law, as recorded by a GM tube, when a large fraction of the beta particles is lost from the beam.

A low attenuation coefficient corresponds to high penetrating power and vice versa.

Avogadro constant is the number of carbon atoms in 0.012 kg of carbon–12. It equals, by definition, the number of elementary entities in a mole. The symbol used is L and its magnitude is 6.022×10^{23} mol⁻¹.

What other subjects are you studying?

A–Zs cover 18 different subjects. See the inside back cover for a list of all the titles in the series and how to order.

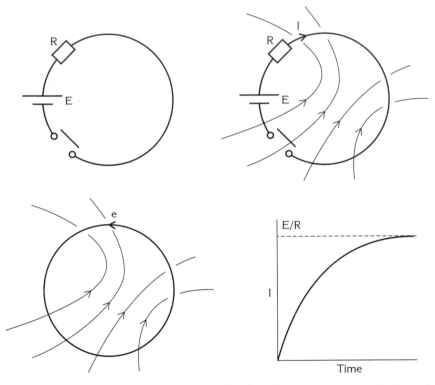

B-field: a three-dimensional space in which the magnitude of the resultant *magnetic field strength B* at any point is mostly non-zero. The zero values of *B* are found at so-called neutral points.

back e.m.f. is an alternative name for *induced e.m.f.*, an alternative which stresses the directional element of the induced e.m.f.

Consider the above set of related diagrams. The first diagram shows a cell of e.m.f. *E* in series with resistance *R* and an open switch S. The second diagram shows the steady current *I* in the circuit and the associated magnetic field some time after the switch has closed. The third diagram shows the growing magnetic field and the induced e.m.f. *e* immediately after the switch S is closed. Theoretically, there are three possibilities:

- the direction of the induced e.m.f. **e** is shown correctly in the third diagram

- the direction of this **e** induced e.m.f. is shown the wrong way round

- there is no induced e.m.f.

Since experiment shows that there is an induced e.m.f., the third possibility is wrong. The second possibility would imply that the induced e.m.f. and the supply voltage were in the same direction. The current in the circuit would then exceed the equilibrium value *I* and power dissipation in the resistor would exceed the power supplied by the battery. This would offend against the law of conservation of energy and cannot be considered as a possible outcome. The first possibility must be correct and the direction of the induced e.m.f. must oppose the supply voltage. For this reason it is called a back e.m.f.

Notice that the effect of the back e.m.f. is to slow down the growth of the current to its final value. The actual line followed by the graph depends on the resistance of the resistor *R* and on the inductance *L* of the circuit loop. Inductance in an electrical circuit behaves likes mass in a mechanical situation. See *electromagnetic induction (laws of)*.

background radiation is the recorded count rate in the absence of any radiation sources which are about to be monitored. It is the count rate linked to the environment and its two principal sources are: (a) rocks and building materials and (b) cosmic radiation. The strength of the former source depends on the immediate environment. The strength of the latter source is roughly one particle per square centimetre per second.

The measured count rate near a radioactive source exceeds the true count rate by the background radiation count rate. To ignore the background radiation count rate would be to introduce a systematic error into the experiment. It is good practice to measure the background radiation count rate several times during an experiment and then to take an average.

bar: a unit of pressure equal to 0.1 MPa. Atmospheric pressure rises and falls by a few per cent on either side of 1000 mbar. Meteorologists use the millibar because they can easily be converted into pascals (1 mbar = 100 Pa) and used in analytical equations with SI units, whereas pressures in units of the standard atmosphere are more awkward. (See *standard pressure*.)

barometer: see *mercury barometer*.

baryons are *hadrons* made up of three *quarks* to give integral or zero charge, baryon number 1 or –1 and spin ½. Masses are rarely much higher than 1 GeV/c^2. The proton and neutron masses are 0.938 GeV/c^2 and 0.939 GeV/c^2, respectively. Although there are lots of possible baryons that could be formed, given that there are 12 different quarks and antiquarks, all baryons other than *protons* and *neutrons* are short lived. Protons and neutrons are the only baryons that matter at A level.

Protons and neutrons are stable within the *nucleus*. The neutron is unstable outside the nucleus: it decays with *half-life* of about a quarter of an hour into a proton, an *electron* and an antineutrino. Other baryons decay with half-lives of around 10^{-10} s if the strong interaction is responsible for the decay or with a half life of around 10^{-24} s if the weak interaction is responsible for the decay. See *Feynman diagrams* for the beta decays of protons and neutrons.

Protons have charge +e and the quark mix uud; neutrons have charge zero and the quark mix udd. Protons and neutrons are long lived partly because the u and d quarks are themselves stable. Beta decay occurs when the weak interaction prompts a d quark to change into a u quark within a neutron or (less likely) a u quark into a d quark within a proton.

The main properties of eight important baryons are summarised in the following table. The quantum numbers of the corresponding antiparticles are not shown but they have the same magnitudes with the opposite signs. Particles are built from quarks and their antiparticles are built from the corresponding antiquarks.

Name	Quantum numbers			Quark content
	Q	B	S	
proton	+1	1	0	uud
neutron	0	1	0	udd
lambda	0	1	−1	uds
sigma plus	+1	1	−1	uus
sigma minus	−1	1	−1	dds
sigma zero	0	1	−1	uds
xi minus	−1	1	−2	dss
xi zero	0	1	−2	uss

base quantity: one of the seven physical quantities corresponding to the seven *base units* in SI. They are length, mass, time, electric current, temperature difference, amount of substance and luminous intensity.

base unit: one of the seven base units of SI whose magnitudes are defined without reference to other units (although they may refer to one another). They are as follows:

- metre, unit of length, symbol m: the distance travelled by light in vacuum during a time interval of 1/299 792 458 of a second

- kilogram, unit of mass, symbol kg: the mass of an international prototype held at Sèvres in France.

- second, unit of time, symbol s: the duration of 9 192 631 770 periods of the radiation corresponding to the transition between the two hyperfine levels of the ground state of the caesium-133 atom.

- ampere, unit of electric current, symbol A: the constant current which, by flowing in two parallel and infinitely long straight conductors of negligible cross-section and held one metre apart in vacuum, sets up a force on these conductors equal to 2×10^{-7} newton per metre of length.

- kelvin, unit of thermodynamic temperature difference, symbol K: the thermodynamic temperature difference between absolute thermodynamic zero and the triple point of water divided by 273.16.

- mole, unit of amount of substance, symbol mol: the amount of substance which contains as many elementary entities as there are atoms of carbon in 0.012 kg of carbon–12.

- candela, unit of luminous intensity, symbol cd: the luminous intensity in a given direction of a source which emits monochromatic radiation of frequency 540×10^{12} Hz and whose radiant intensity in that direction is $\frac{1}{683}$ watt per steradian.

These seven units are the complete set. You are unlikely to need the candela.

(See *base quantity, prefixes used with SI units, metre, kilogram, second, ampere, kelvin, mole, candela.*)

beams are long bars of *rigid* material used for bridging gaps. A beam is possibly the simplest example of a structure. The exaggerated sagging of the beam in the diagram high-lights that, when in place, the lower half of a beam is in tension whilst the upper half is in compression. The tension along the lower half of the beam can only be withstood by a material which is strong and the deflection will be small only for a material with a high Young's modulus. This is why steel is favoured for beams. Where concrete is used, the beam is reinforced with steel rods along its lower half. Concrete beams are even stronger if tensioned steel rods are used to pre-stress the concrete. A low value for the bending implies high stiffness. (See *composite materials, strength*.)

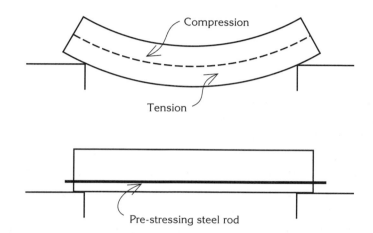

becquerel: the SI unit of *activity*, the symbol used is Bq. A radioactive source of activity (strength) 1 Bq has an activity of 1 disintegration per second. (See *frequency, fundamental law of radioactive decay*.)

Bernouilli's principle states that the *pressure p* in a fluid, in conditions of streamline flow, satisfies the equation:

$$p + \tfrac{1}{2}\rho v^2 + \rho g h = \text{constant}$$

Streamline flow is the smooth flow of a fluid without any eddies due to turbulence. It is destroyed when a fluid flows along a pipe or a canal too fast.

Notice that all the quantities in the equation are evaluated at a point; p is thrust per unit area at a point, ρ is mass per unit volume at a point, v is the rate of change of displacement at a point, g is the gravitational field strength at a point and h is the vertical distance of the point from a reference level.

The middle term in the expression, $\tfrac{1}{2}\rho v^2$, is called the dynamic pressure. The static pressure, $p + \rho g h$, is the sum of an intrinsic component p and a gravitational potential energy component $\rho g h$. The intrinsic component matters more in gases, while the gravita-tional potential energy component is often the significant component in moving liquid masses.

In key applications, the change in level h is usually too small to be relevant and an increase in speed brings about a drop in pressure. Three applications of the principle are illustrated in the diagram on page 22.

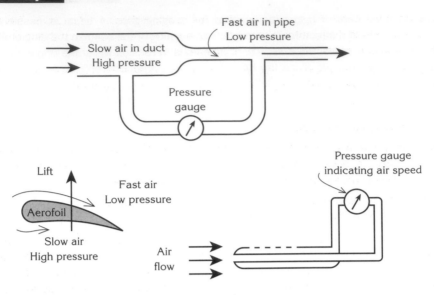

The pressure gauge in the first diagram records the higher pressure in the duct compared with the pipe; the air is moving faster within the pipe than within the duct. This arrangement is the basis of carburettor design.

The second example is the aerofoil. It is so shaped that air flows faster over one surface than the other. There is a drop in pressure which gives a net force(or *lift*) on the aerofoil from the low speed side towards the high speed side. This is the principle of the *aeroplane* wing but it explains the action, too, of sails on *sailing boats* and their rudders.

The third example is the pitot-static tube. The tube is mounted parallel to streamlines in a moving fluid. The fluid runs past the holes in the outer tube and reduces the pressure. The inner tube has a hole facing the current and catches its full force. The reading on the pressure gauge can be used to calculate the speed of the fluid. (See *aeroplane*, *motor boats*, *sailing boats*.)

beta decay is the spontaneous emission of a beta particle from a nucleus. If the beta particle is negatively charged, it is an electron and an antineutrino is expelled from the nucleus at the same time. If the beta particle is positively charged, it is a positron and a neutrino is expelled at the same time.

The general equation for β^- decay is

$$_{Z}^{A}X \longrightarrow\ _{Z+1}^{A}Y +\ _{-1}^{0}e + \bar{\nu} + \text{energy}$$

X is the parent nuclide, Y is the daughter nuclide and $\bar{\nu}$ represents an antineutrino. The mass of the antineutrino, if it has any, is too small to measure and the energy released comes from the mass excess of X compared with [Y + e].

If there were no antineutrinos, all beta particles from a pure sources should have the same energy – as is the case with alpha particles. But beta particles are known to have a spread of energies up to a maximum value decided by the mass difference between the two sides of the decay expression, (see *beta radiation*). The neutrino was 'invented' to allow for this spread of energies while retaining the conservation of energy law. Later still, when it was recognised that electrons and neutrinos are both *leptons* and that lepton number must be conserved in the decay process, the neutrino was re-labelled as an antineutrino. The origin of beta decay is the decay, within the nucleus, of a neutron into a proton with the emission of an electron and an antineutrino (See *Feynman diagrams*).

If a beta particle is positively charged, it is a positron, in which case a neutrino is expelled at the same time. For β^+ or positron decay, the decay equation is written:

$$_Z^A X \longrightarrow \ _{Z-1}^A Y + \ _{+1}^0 e + \nu$$

where ν represents a neutrino.

β^+ decay is a rare event, mostly because any nuclei which are likely to decay this way are much more likely to decay by *K-capture*. See also *beta radiation, N–Z curve for stable nuclei, four fundamental interactions* and *proton decay and neutron decay*.

beta particle: an *electron* or a *positron* emitted from the nucleus of an atom during radioactive decay. A flux of beta particles is called *beta radiation*.

beta radiation consists primarily of electrons emitted from radioactive nuclei but the term also covers streams of positrons (positively charged electrons). The properties of a beam of *beta particles* seem not to be quite the same as those of a beam of electrons in a cathode ray tube, for instance. This is because electrons in a cathode ray or fine beam tube are accelerated through a few thousand volts, whereas electrons expelled from a nucleus during *beta decay* have about a thousand times as much energy. Beta particles are deflected by magnetic fields, but a much stronger field is needed to give even a modest deflection with the lower speed electron beam. Beta particles penetrate a thin aluminium sheet where a low speed electron beam would do no more than warm its surface. The high penetrating power makes beta radiation a severe health hazard to any animal exposed to even a modest dose. Deep body damage is possible.

Alpha particles emitted from any one radioactive nuclide all have much the same energy and consequently the same range in air (see diagram below). But the diagram shows how beta particles are emitted with a range of energies from zero (improbable) to a maximum value (also improbable). The reason is that in beta decay a second particle is emitted from the nucleus (the antineutrino) and the available energy is shared more or less randomly between the two particles.

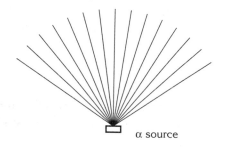

Range in air of alpha particles
from a pure alpha emitter

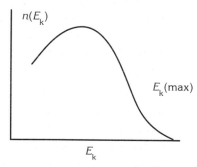

Number of beta particles with
energy E_k, $n(E_k)$, against
energy when emitted E_k

Beta radiation can be detected and monitored efficiently with a *GM tube* attached to a scaler.

binding energy is the energy released when a heavy nucleus is formed (in the imagination) spontaneously from its constituent particles. A simple example is offered by the alpha particle. The masses of the relevant particles are listed below in kilograms and as multiples of the *unified atomic mass constant*, u.

$$\text{proton mass} = 1.672\ 623 \times 10^{-27}\ \text{kg} = 1.007\ 276\ \text{u}$$
$$\text{neutron mass} = 1.674\ 928 \times 10^{-27}\ \text{kg} = 1.008\ 665\ \text{u}$$
$$\text{alpha particle} = 6.644\ 46 \times 10^{-27}\ \text{kg} = 4.001\ 5\ \text{u}$$

The total mass of two protons and two neutrons is $6.695\ 103 \times 10^{-27}$ kg or $4.031\ 883$ u.

The '*mass defect*' for the alpha particle is thus 5.066×10^{-29} kg or 0.0305 u.

This 'lost mass' has energy equivalent ($\Delta m.c^2$) equal to 4.553×10^{-12} J or 28.4 MeV and this is the binding energy of the alpha particle. The binding energy of a nucleus is the work which would have to be done to separate it into its constituent particles. The *binding energy per nucleon* for the alpha particle is 7.10 MeV per nucleon.

binding energy per nucleon for an atomic nucleus equals the *binding energy* for the nucleus divided by the number of *nucleons*. The graph below shows in broad terms how binding energy per nucleon varies with atomic number Z.

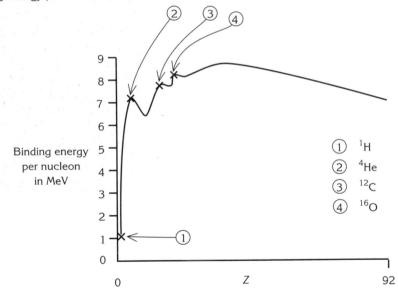

The smoothed out graph peaks near iron–56. Since binding energy is energy released on formation, not stored energy, it is clear that light nuclei can release energy by fusing (going to a state of higher atomic number and higher binding energy) while heavy nuclei can release energy by breaking up (fission).

biofuels are materials such as wood which store solar energy recently harnessed by living matter. (See *energy sources*.)

black body radiation and total radiation mean the same thing. (See *total radiation curve*.)

black hole: the end state of a main sequence star whose mass exceeds about $20 M_{\text{sun}}$. The gravitational field near the surface of a black hole is so strong that nothing, not even light, can escape. See *end states of stars*.

Boltzmann constant k equals the ratio R/L where R is the gas constant per mole and L is the Avogadro constant. L equals the number of molecules in a mole and k is the gas constant per molecule of gas. Its symbol is k and the unit is J K^{-1}. The value of the Boltzmann constant is 1.38×10^{-23} J K^{-1}.

The phrase 'gas constant per molecule' means the increase in the gas constant for a specimen of gas, per molecule added to the specimen. It does not mean the gas constant for an isolated molecule.

The formal definition of the Boltzmann constant is rooted in thermodynamics and outside the scope of this book. An important result is however, the relationship between absolute temperature and molecular kinetic energy.

For a monatomic gas:

$$\tfrac{1}{2}m<c^2> = \tfrac{3}{2}kT$$

where m is the molecular mass and $<c^2>$ is the mean square speed of the molecules. This result is used to import the temperature variable into kinetic theory. (See *kinetic theory of gases*.)

Boyle's law (1660) states that the pressure of a fixed mass of gas at constant temperature and in an equilibrium state is inversely proportional to its volume. It can be stated algebraically:

$$\text{provided that } m \text{ and } T \text{ are constant, } p \propto \frac{1}{V} \quad \text{or} \quad p_1 V_1 = p_2 V_2 = \text{constant}$$

Although the law is easily derived from the general gas equation, it is an empirical law and was discovered by experiment. An account of a student experiment to verify the law should include a diagram, a list of the measurements and an explanation of how the law would be checked.

Experiment to verify Boyle's law

Apparatus

Set up apparatus similar to that shown in the diagram. There is a container, in which a fixed mass of gas is trapped, immersed in a vessel filled with water whose temperature can be checked. There is a scale from which the volume of the gas can be read. There is a means of changing the pressure of the gas and a pressure gauge to measure its new value. If the pressure gauge measures *gauge pressure*, then the atmospheric pressure at the time of the experiment must be measured separately.

Measurements

Once the temperature of the gas in the container has settled down, the initial volume should be read from the volume scale and the initial pressure should be read from the pressure gauge. Using the foot-pump, the pressure should be increased in about six equal stages up to a maximum value. At each stage, the air temperature is given time to settle at its initial value and the volume and pressure values are noted. If the pressure gauge reads the gauge pressure, then the local value for atmospheric pressure should be added to all the pressure gauge values.

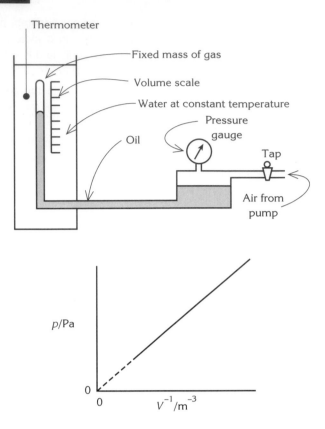

Conclusion

A graph is drawn of pressure against 1/volume. If the graph is a straight line passing through the origin, then it verifies Boyle's law at this one temperature and for the range of pressures used in the experiment.

(See *ideal gas equation, Charles' law, pressure law, gauge pressure*.)

breaking stress is the tensile *stress* at which a material fractures. Alternative terms are *strength* (which may relate to the force or the tensile stress) and *ultimate tensile stress*.

brittleness is the quality present in some materials (particularly glasses and ceramics) of breaking without warning. Brittleness implies an absence of any form of *plastic deformation*, or flow to warn of an impending break up. (See also *stress–strain curves*.)

Brownian motion is the random and erratic motion of small particles suspended in a fluid. The suspension may be cigarette smoke in air or pollen particles in water. The particles must be small enough to move about when fewer atoms hit it on one side than on the opposite side but large enough to be visible.

The microscope objective collects light over a large angle. In this system the microscope acts more like an image intensifier than a magnifier. It brightens the spots of light and spreads them out, but the particles themselves remain invisible. If the magnification is too great the depth of field is too shallow and particles are seen coming in and out of focus instead of moving around at random.

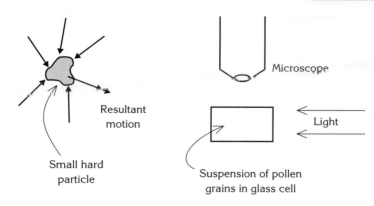

Resultant motion

Small hard particle

Microscope

Light

Suspension of pollen grains in glass cell

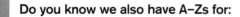

During the nineteenth century Brownian motion was little more than a novelty which arguably demonstrated the atomic theory of matter. It became a significant phenomenon early in the twentieth century when Einstein showed how measurements of the gross movements of particles after numerous collisions could lead directly to an independent calculation of the *Avogadro constant*.

Do you know we also have A–Zs for:

- **Chemistry**
- **Biology**
- **Mathematics**
- **ICT & Computing?**

Ask for them in your local bookshop or see the inside back cover for ordering details.

calibration is the procedure for marking out and interpreting the numerical scale on an instrument. An example would be converting the voltage scale on a digital voltmeter to a temperature scale to match the temperatures indicated by a particular *thermocouple*. Each number of the completed scale must be multiplied by a unit (e.g. kelvin) and possibly by a multiplier (e.g. 10^3). See *measurement*.

capacitance, C, is the ratio of electric charge, Q, on a conductor to its electric potential, V, assuming that the electric potential of the conductor is zero when the charge is zero. Capacitance is defined by the word equation:

$$\text{capacitance} = \frac{\text{charge}}{\text{potential}}$$

$$C = \frac{Q}{V}$$

The same word equation defines the SI unit of capacitance, the farad, F:

$$\text{farad} = \frac{\text{coulomb}}{\text{volt}}$$

This definition applies to an isolated conductor. It does not apply to a charged insulator; an insulator cannot have a potential because charge cannot flow across the insulator to equalise the potential.

The usual situation is to build a capacitor from conductors placed close together. They are close enough for lines of force which begin on the positively charged conductor to all end on the negatively charged conductor. Capacitance equals the charge stored on one of the conductors divided by the potential difference across the two conductors.

$$\text{capacitance} = \frac{\text{charge stored on one plate}}{\text{potential difference between plates}}$$

(See *capacitance (measurement of)*, *energy stored in a capacitor*, *parallel plate capacitor*.)

capacitance (measurement of): involves measuring the charge stored in a capacitor when the potential difference between the plates is known. In order to measure capacitance directly, we need some way of measuring the charge on a capacitor. One way of doing this is to measure the charge flowing into the capacitor by multiplying a constant charging current by the time of current flow.

First, consider some order-of-magnitude calculations. Imagine a 1000 µF capacitor (i.e. 1 mF) with 5 V across it. The charge stored is 5 mC. This is the charge transferred by a current of 50 µA in 100 seconds. And a 5 V supply will drive a current of 50 µA through a 100 kΩ resistor.

Experiment: to measure capacitance

Apparatus

Set up the circuit in the diagram below. The variable resistor has a maximum value of 100 kΩ. The initial charging current from the 5 V power supply will be 50 μA and the current in the digital voltmeter will be no more than a very small fraction of 50 μA.

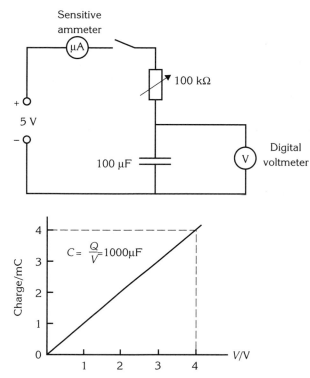

Measurements

Switch on the circuit and use the variable resistor in conjunction with the microam-meter to keep the current, as far as possible, at a constant value of 50 μA until the voltmeter reading stops rising. Note the voltmeter readings at 10-second intervals, starting from the moment the circuit is switched on. Repeat the experiment three or four times and average the corresponding voltmeter readings. The charge on the capacitor increases by 1 mC in 20 seconds for a constant current of 50 μA (1 mC equals 50 μA times 20 seconds). You should manage eight or nine voltage values.

Conclusion

With about eight pairs of charge and voltage measurements, a graph can be drawn of charge against voltage. The gradient is the capacitance.

Comment

The most serious source of error is the current: how skilled are you at keeping its value constant. This will be a random error – and random errors can be reduced by averaging several independent measurements. Also, you will have the opportunity to practice a few times before the serious measurements begin.

This experiment serves other functions. It shows that the potential difference across the capacitor is proportional to the charge stored. It could be used to measure the average value of a small current by measuring the charge transferred in a known time to a known capacitance. Again, using a known capacitance, the circuit could be used to verify the relation between charge and current (charge transferred = current × time).

capacitive reactance is the ratio of peak voltage to peak current for a capacitor. One of the complications of AC circuits is that voltage and current peak at different times. We shall see how this comes about.

The diagram below shows a capacitor of capacitance C connected across a power supply with output voltage V equal to V_0 sin $2\pi ft$. or V_0 sin ωt. The *angular frequency ω* radians per second equals $2\pi f$. The frequency in cycles per second is f.

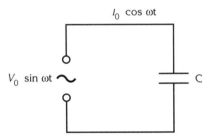

The behaviour of the capacitor is governed by the equation:

$$Q = CV$$

where Q is the charge stored in the capacitor at an instant when the voltage across it is V. But

$$V = V_0 \sin \omega t$$

and

$$I = \frac{dQ}{dt}$$

If we put

$$Q = CV = CV_0 \sin \omega t$$

then

$$I = \frac{dQ}{dt} = \frac{d}{dt}(CV_0 \sin \omega t)$$

$$= \omega CV_0 \cos \omega t$$

$$= I_0 \cos \omega t$$

where

$$I_0 = \omega CV_0$$

Compare this with Ohm's law:

$$V = IR \qquad V_0 = \left(\frac{1}{\omega C}\right)I_0$$

In Ohm's law, the resistance R holds the balance between current and voltage. For the capacitor, $1/\omega C$ holds the balance between peak voltage and peak current but these two quantities are not in phase; the current peaks before the voltage. The term $1/\omega C$ is called the reactance of the capacitor. The symbol for capacitive reactance is X_c and the SI unit for reactance is the ohm.

The best way of seeing what all this means is by working carefully through an example.

Worked example

Let the power supply have a peak output V_0 of 5.0 V, let its frequency f be 50 Hz (ω equals $2\pi f$ or 314 rad s^{-1}) and let C equal 10 µF. Start by finding the reactance of the capacitor:

$$X_C = \frac{1}{\omega C} = \frac{1}{(314 \text{ rad s}^{-1})(10 \times 10^{-6} \text{ F})} = 0.318 \text{ k}\Omega$$

With peak voltage at 5.0 V, the peak current is 5/318 or 15.7 mA (16 mA to 2 s.f.). Equations for voltage and current can now be written out in detail and plotted on a graph:

$$V = 5.0 \sin 314t$$

$$I = 0.016 \cos 314t$$

314 is a large number and this is why the time axis on the following graph is in milliseconds. Notice that the current axis is calibrated on the left-hand side of the graph and the voltage axis calibrated on the right-hand side.

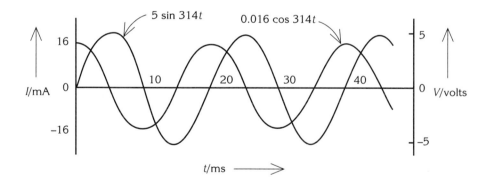

The time axis is drawn with time increasing towards the right. The current in the capacitor peaks a quarter of a cycle earlier than the voltage. This explains the awkward phrase 'the current leads the voltage'. The circuit would need to be switched on for many cycles before the current would settle down at the values shown in the graph.

(See *frequency, phase, reactance, capacitors in a.c. circuits, series resistance and capacitance, inductive reactance, series resistance and inductance, resonance in AC circuits*.)

capacitor: a device for storing electric charge. The *capacitance* of a capacitor is defined by the following word equation which can also be used to define the unit of capacitance, the *farad*.

$$\text{capacitance} = \frac{\text{charge stored}}{\text{potential difference}}$$

$$C = \frac{Q}{V}$$

$$\text{farad} = \frac{\text{coulomb}}{\text{volt}}$$

A capacitor usually has two plates which are insulated from one another and two terminals. The potential difference V is then the voltage between the terminals. Should there be one terminal only, as would be the case with the capacitance of an isolated metal spoon, then the relevant potential difference would be the voltage between the spoon and an earthed terminal somewhere. See *parallel plate capacitor*.

capacitor limitations are primarily peak voltage, peak frequency, leakage current, stability (i.e. retaining constant characteristics) and size. These are the parameters which should be considered when choosing a capacitor.

Large capacitance (in excess of 1.0 µF) is almost always achieved with electrolytic capacitors. The dielectric between the plates is a polarised electrolyte which, because it can be made very thin, gives large capacitance in a small volume. There are two disadvantages however with electrolytic types. First, the capacitor has a positive and negative plate and cannot withstand an alternating voltage. Secondly, electrolytic capacitors are slow-acting and can only be used with modest frequencies (a few kHz). Their stability is never perfect but this tends not to matter in typical applications such as smoothing circuits.

Other capacitors with mica or paper dielectrics can work at much higher frequencies and with alternating voltages but their capacitance values are much smaller.

capacitor size increases with capacitance and peak voltage (with peak charge stored). (See *capacitor limitations*.)

capacitors in AC circuits follow the same rule as capacitors in DC circuits. Charge stored is proportional to voltage, $Q = CV$. If C is constant and V alternates, then Q will alternate in phase with V.

But the current in the circuit equals the rate of increase of the charge stored on the capacitor (it is the gradient of the charge–time graph).

The diagram below shows a capacitor in series with an AC supply and graphs of voltage, charge and current against time. Notice the use the subscript $_0$ to indicate a peak value.

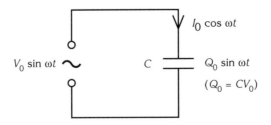

The output of the power supply or signal generator is the starting point for any calculation. The instantaneous charge stored equals capacitance × instantaneous voltage. Current equals rate of change of charge stored. If you do not understand calculus, just accept the answers are correct.

$$V = V_0 \sin \omega t$$

$$Q = CV = CV_0 \sin \omega t = Q_0 \sin \omega t$$

$$I = \frac{dQ}{dt} = \frac{d}{dt}(CV_0 \sin \omega t) = \omega CV_0 \cos \omega t = I_0 \cos \omega t$$

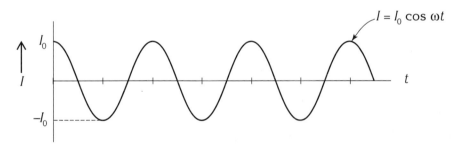

$V = V_0 \sin \omega t$

$Q = Q_0 \sin \omega t$

$I = I_0 \cos \omega t$

Compare the relationship between the magnitudes of peak current and peak voltage (ignoring the fact that they occur at different times) with the corresponding DC equation for resistance:

$$I_0 = \omega CV_0 \quad \text{or} \quad V_0 = \left(\frac{1}{\omega C}\right)I_0$$

$$I = \frac{1}{R}V \quad \text{or} \quad V = RI$$

The quantity $1/\omega C$ holds the balance between peak voltage and peak current for a capacitor in an AC circuit. It is called the reactance of the capacitor.

Worked example

The peak voltage for a signal generator is 7.1 V. The signal generator is connected across a 0.10 µF capacitor and the frequency is set at 2.5 kHz. Calculate the peak current.

Using the relationship $\omega = 2\pi f$, we have ω equals 15.7×10^3 rad s^{-1}.

$$I_0 = \omega C V_0$$
$$= (15.7 \times 10^3)(0.1 \times 10^{-6})(7.1) = 11 \text{ mA}$$

(See *series resistance and capacitance, r.m.s. values, resonance in AC circuits, AC symbols*.)

capacitors in parallel are connected side-by-side as shown in the diagram below. A single capacitance C will be equivalent to the parallel set, provided that the charge Q it stores for the same potential difference V, equals the sum of the charges stored on the individual capacitors in the set. For equivalence, then:

$$Q = Q_1 + Q_2 + Q_3$$

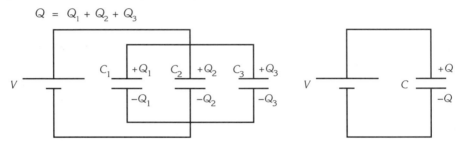

Remembering that capacitance is defined by the equation $Q = CV$, we have:

$$CV = C_1 V + C_2 V + C_3 V$$
$$C = C_1 + C_2 + C_3$$

(See *capacitors in series, energy stored in a capacitor*.)

capacitors in series are capacitors joined end-to-end as shown in the diagram below. The equivalent capacitor, shown isolated on the right, must store the same charge when the voltage across it is the same as the voltage across the series set, so:

$$V = V_1 + V_2 + V_3$$

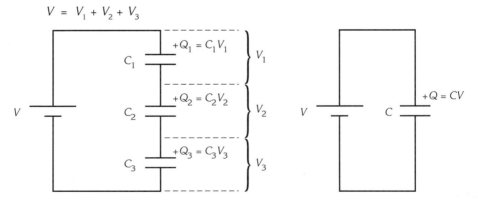

The same charge Q must be stored on each of the capacitors in the series set otherwise the adjacent positive and negative charges could not cancel out on discharge. The isolated capacitor must also carry charge Q if it is equivalent to the series set. Remembering that capacitance is defined by the equation $Q = CV$, we have:

$$V = V_1 + V_2 + V_3$$

$$\frac{Q}{C} = \frac{Q}{C_1} + \frac{Q}{C_2} + \frac{Q}{C_3}$$

$$\frac{1}{C} = \frac{1}{C_1} + \frac{1}{C_2} + \frac{1}{C_3}$$

One consequence of this relationship is that the equivalent capacitance C of a series set is always smaller than the smallest capacitance in the set. (See *capacitors in parallel, energy stored in a capacitor*.)

Celsius temperature scale is defined from the *absolute temperature* scale by the relationship:

$$\theta/°C = T / K - 273.15$$

It is another way of choosing the numbers for the absolute temperature scale. Since the ice point is zero degrees Celsius and the steam point one hundred degrees Celsius, it is a *centigrade temperature scale*.

centigrade temperature scale is any temperature scale which has its *ice point* at 0°C and its *steam point* at 100°C. There is no limit to the number of possible centigrade scales because there are numerous ways of defining how temperatures other than the ice and steam point are to be measured. The *Celsius temperature scale* uses the *absolute temperature scale* to define these other temperatures.

centre of gravity is the point through which the weight of a rigid body acts. No matter how much the body tumbles or turns, it is a fixed point relative to the body.

If a rigid body is imagined to consist of innumerable tiny masses, the centre of gravity is the point about which the sum of the moments of the weights of all these tiny masses is zero. It is the point about which the body will balance.

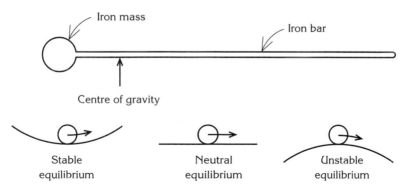

A rigid body resting on a surface is in stable equilibrium if any small rotation raises the centre of gravity. It is in unstable equilibrium if such a rotation lowers the centre of gravity. It is in neutral equilibrium if such a rotation neither raises nor lowers the centre of gravity. The three cases are illustrated in the diagram above.

The centre of gravity of a rigid body coincides with its *centre of mass* in a uniform gravitational field.

centre of mass of a rigid body is the point through which any resultant force must act if it is to accelerate the body without causing it to rotate. A terrestrial body would have to be very large for the centre of mass and the *centre of gravity* to be in two different places.

centrifugal force is the force which forms a 'Newton's *third law* pair' with the *centripetal force* applied to a body undergoing circular motion. Centripetal force acts on the body undergoing circular motion therefore the centrifugal force must act on some other body. In a spin drier, for instance, the centripetal force is a normal contact force acting on the clothes and holding them in their circular path; the centrifugal force is the normal contact force exerted by the clothes outwards on the spinning container. It is the centrifugal force that acts on the surroundings, not on the spinning body. The topic is usually avoided in A level physics because the term 'centrifugal force' has a confusing history.

centripetal acceleration is the *acceleration* of a body which travels along a circular path at constant speed. Its direction is towards the centre of the circle. (See *circular motion, centripetal force*.)

centripetal force is a force which acts at right angles to a body's path and so constrains the body to follow a circular path about the point towards which the centripetal force acts. For the body to continue round the circle it must be moving at just the right *speed* for that radius. For a body of mass m moving at speed v round a circle of radius r:

$$\text{centripetal force} = \frac{mv^2}{r}$$

Notice that this equation does not tell us what kind of force is acting. It may be a gravitational force, an electrical force, the force on a charge moving through a magnetic field, or the tension in a length of string.

Worked example

An Earth satellite. The free-body force diagram shows that only one force acts on the satellite; its weight. So the weight is the resultant force and, in this instance, the centripetal force too. But weight is the only force acting on the body, it is the only entry on the diagram. Do not write resultant force or centripetal force on the diagram; these terms describe what forces do, they are not part of the list of forces acting on the body. Remember too that the weight gets smaller with increasing distance from the Earth's centre.

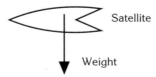

Satellite

Weight

If G is the gravitational constant, M the mass of the Earth, R the radius of the Earth, m the mass of the satellite, v its velocity and h its height above the Earth's surface, then:

$$\frac{mv^2}{(R + h)} = G.\frac{mM}{(R + h)^2}$$

This equation is not saying that the gravitational force equals the centripetal force. That would imply that there are two equal forces. It says that the gravitational force is the centripetal force.

cepheid variable: a *star* whose *luminosity* (i.e. total radiant output power) varies cyclically with a constant period. It happens that the period of one full cycle (typically anywhere between a few hours and a few months) is a function of the average luminosity. It follows that if the period of a cepheid is measured with a clock, then the average luminosity L of that cepheid can be calculated.

The intensity I of the light from the star (radiant energy per square metre arriving at the Earth's surface) can be measured with a telescope. Once both I and L are known for the same cepheid, the distance D of the cepheid can be calculated from the relation $I = L/4\pi D^2$. Since this distance D is much the same as the distance away of the galaxy in which the cepheid is located, the distances of any galaxy that contains a cepheid can be calculated.

chain reaction: a repetitive process in which a reaction product at one stage (a neutron in this example) is a reactant for the next stage. The diagram illustrates the chain reaction for ^{235}U in a *thermal fission reactor*.

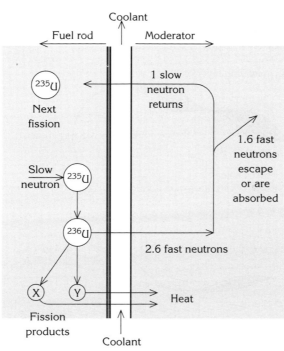

A slow neutron in the fuel rod is absorbed by a uranium-235 nucleus. The uranium-236 nucleus so formed is unstable and divides spontaneously into two *fission* fragments, labelled X and Y, and two or three fast neutrons. The number of neutrons per fission reaction averages out at 2.6.

The total rest mass of the fission products is less than the rest mass of the uranium nucleus and the mass difference appears as the kinetic energy of the fission fragments. Most of this energy is linked to the X and Y nuclei and appears as heat in the fuel rods. This heat is carried

A
B
C
D
E
F
G
H
I
J
K
L
M
N
O
P
Q
R
S
T
U
V
W
X
Y
Z

off by the coolant to power the steam turbines. The fast neutrons escape as far as the moderator where they slow down and enough of them return to the fuel rods to sustain the fission process. Fast neutrons have a low probability of being absorbed by uranium-235. Slow neutrons are sometimes called thermal neutrons because their average kinetic energy E_k matches the reactor temperature with $E_k = \frac{3}{2}kT$.

charge carrier: a charged particle whose movement constitutes electric current. Charge carriers in metals are primarily electrons. In *semiconductors*, they are electrons or holes. Electrons and holes occur in equal numbers in intrinsic semiconductors. There is an excess of electrons in *n*-type semiconductors and an excess of holes in *p*-type semiconductors. It is this asymmetry which makes the p-n junction diode possible.

charge-coupled device (CCD): a rectangular array of light sensitive pixels which forms the image detection sensor in camcorders and some telescopes. For telescopes, the number and distribution of the pixels matches the pixels on a PC monitor.

The first diagram shows one CCD pixel. It consists of three tiny metal flags labelled, respectively a, b and c. One row of pixels corresponds to one row or horizontal line on the PC monitor. Three 'wires' connect all the *a* pixels together, all the *b* pixels together and all the *c* pixels together. The metal flags are mounted on a very thin layer of silicon dioxide which, in turn, covers a thin but stiff sheet of *p*-type semiconductor.

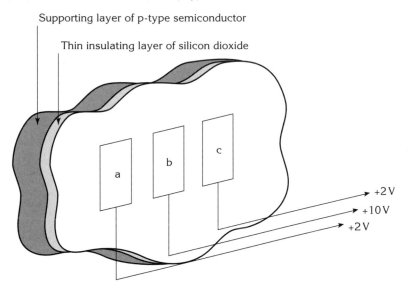

During the detection phase, the *a* and *c* wires are held at +2 V while the *b* wire is held at +10 V. The thin sheet of *p*-type semiconductor faces the incoming light and an image of part of the sky is formed by the telescope on this thin sheet. About 70% of the photons that light up the image create electron–hole pairs within the semiconductor and the electrons from these pairs congregate below the *b* set of metal flags, which are held at the higher positive potential of +10 V. The magnitudes of the charges stored in the pixels form an accurate record of the light intensity distribution falling on the pixels.

To collect the picture for display on a monitor or for storage in memory, the light is blocked off and the potentials of the *a*, *b* and *c* sets of metal flags change in sequence in the manner shown in the diagram. The *c* set changes to +10 V, with the *a* and *b* sets both at

+2 V; the individual charges stored in the semiconductor below the b flags step to the right. Next, the a set switches to +10 V with the b and c sets both at +2 V. The charges step to the right again. All these charge concentrations are swept in sequence to collectors where the charge amplitudes are measured and stored. With one collector for each row, a picture is soon built up.

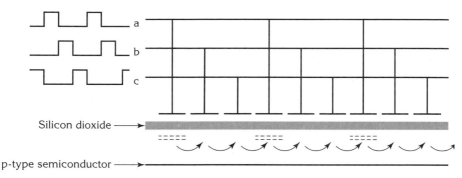

The advantage of the CCD detector is the 70% photon-detection efficiency compared with less than 5% for photography. Much fainter star systems can be observed and examined. Photographs still have the advantage in that their grain size can be much smaller than the pixel size. Where resolution is of the essence, long-exposure photographic recording still wins.

charge–time graph: see *current–time graph*.

charge-to-mass ratio is the electric charge of a charged particle in coulombs, divided by its mass in kilograms. It is little more than a useful number to give insights, and perhaps this is why it rarely has a symbol of its own and is mostly written out as a ratio. It is called 'specific charge' where the word 'specific' has the meaning 'per unit mass'. Specific charge for the electron is about 2000 times that of the *proton* and about 8000 times that of an *alpha particle*. The higher value for electrons explains why *beta radiation* is so strongly deflected by magnetic fields and why alpha radiation is so weakly deflected.

The ratio e/m_e for the electron works out at about 1.76×10^{11} C kg^{-1}. Electrical forces dominate. One way of thinking about the oil drops used by Millikan to measure the electron charge is as an artificial way of reducing the specific charge by increasing the mass without changing the charge. The charged oil drops used in the *Millikan oil drop experiment* had so low a specific charge that the electrical force on a drop was of the same order of magnitude as the gravitational force.

Charge carriers in vacuum include electrons, alpha and beta radiation and ions.

Charles' law (1787) states that the volume of a fixed mass of gas at constant pressure is proportional to its *absolute temperature*. It can be stated in algebraic form:

$$V \propto T \qquad \text{or} \qquad \frac{V_1}{T_1} = \frac{V_2}{T_2} = \text{constant}$$

Although Charles' law is easily derived from the general gas equation, it is an empirical law and was discovered by experiment. The history of its discovery is complicated by the history of temperature scales. Statements of the law based on the Celsius scale are awkward. An account of a student experiment to verify the law should include a diagram, a list of measurements and an explanation of how you would check the law.

Experiment: to verify Charles' law

Apparatus

Set up apparatus similar to that shown in the diagram. A fixed mass of dry air is trapped in narrow glass tube, a capillary tube, and sealed at the open end by a pellet of concentrated sulphuric acid. The acid keeps the air dry. A scale alongside the tube gives the length of the air column and this can be used as a volume measurement. The thermometer attached along the side of the tube gives the temperature of the air. The tube is immersed in water and the water is heated with an electric heating element and kept stirred.

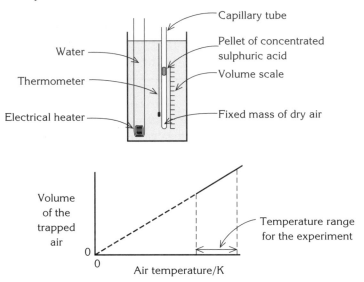

Measurements

When the apparatus has settled down at room temperature, the gas volume and the gas temperature are recorded. The heater is switched on for a while and then switched off. The apparatus is given time to settle down at the new temperature. The new temperature and the new volume are recorded. This procedure is repeated in stages up to about 90°C. Temperatures in degrees Celsius are converted to the absolute scale by adding 273 to each temperature.

Conclusion

A graph is drawn of volume (in the arbitrary unit of volume per unit length of glass tube) against absolute temperature. If the graph is a straight line passing through the origin, then it verifies Charles' law for the value of the ambient atmospheric pressure and for the range of temperatures used in the experiment.

(See *ideal gas equation, Boyle's law, pressure law*)

chromatic aberration is the failure of a glass lens to bring all the colours in a parallel beam of white light to the same focus. The refractive index of glass for light at the blue end of the spectrum is marginally greater than the refractive index for light at the red end. Hence the focal lengths are slightly different and white images are weakly coloured at the edges.

This gives an advantage to reflecting telescopes when larger aperture size, leading to greater light-collecting power and higher resolution, is what an astronomer is looking for.

circular motion is the motion of a body which follows a circular path at constant speed.

In the diagram below the speed of the body is constant, but its direction changes continuously, so the velocity is not constant. With its velocity changing, the body must be accelerating. This acceleration has direction from the body towards the centre of the circle; it is called *centripetal acceleration*. For a body moving round a circle of radius r at constant speed v:

$$\text{centripetal acceleration} = \frac{v^2}{r}$$

Acceleration implies the presence of a force, which is the centripetal force and is just strong enough to hold the body in its circular path. Notice that at A level we only deal with the steady state situation for which the above equation applies. We do not worry about the complex forces which once acted on the body to get it moving along the circular path at just the right speed.

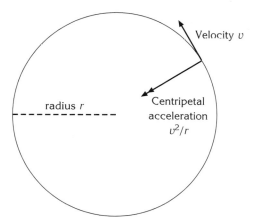

The direction of the centripetal force can be worked out from energy considerations. The speed of the mass is constant and its kinetic energy is constant, so the centripetal force does no work and must be at right angles to the motion. The centripetal force direction must be in the plane of the circle, otherwise the mass would move out of this plane. These two ideas together put the direction of the centripetal force parallel to the radius. It must act towards the centre, otherwise the circular path would be turning the other way.

cleaving is the breaking of a crystal into two parts by splitting it apart between two adjacent and parallel crystal planes. The new surfaces are clean, smooth, uniform and shiny.

coherence implies a permanent phase match between the waves emitted by two separate sources. The rise and fall of one wave perfectly echoes the rise and fall of the other. Coherence is a necessary condition for the two waves to interfere in any way. Should the two waves overlap at some point further from one source than the other, then the two waves may arrive at the point in phase, out of phase or half and half. This is what makes *interference* possible. They will interfere constructively if in phase, the two waves enhancing one another. They will interfere destructively if out of phase, the two waves cancelling one another. If the phase difference is somewhere between these two cases, then enhancement will be partial.

Two light beams can only be coherent if they come from the same source. Light from one electric lamp is altogether different from light from another, apparently identical, lamp. Together, they will brighten a room but there will be no interference pattern. The two beams of light are not coherent.

collision is a situation in which *linear momentum* is transferred from one body to another. Contact is not essential. Repelling magnets can collide and separate without touching.

colour is a notional quality associated with *quarks*. Any one of the twelve kinds of quark can take on one of three values or 'colours'. It is a way of allowing three quarks to occupy similar states without violating the Pauli exclusion principle.

companding is a way of reducing the distortion of an *analogue signal* during transmission. High amplitude components in a signal are more subject to distortion and noise during transmission than low amplitude components. The upper halves of the higher voltage components of the analogue signal are reduced by, say, 70% of their original size before transmission. The lower voltage components are left alone. This part of the process is called compression. The transmitted signal picks up less noise and suffers less distortion than would happen without this compression. The compression is reversed by a corresponding expansion of the larger voltage components after signal reception. The complete process, compression before transmission and expansion after transmission is called companding. The process carries the additional advantage of spreading the lower amplitude values across a larger number of digital values, thereby reducing somewhat the error introduced when the signal is quantised. See *analogue-to-digital conversion*.

composite materials are built up from two or more simpler materials to provide a structured material with enhanced or modified properties. The structure can be ordered, as in plywood, or random, as in concrete. The strength of a concrete *beam* can be further enhanced by including steel rods within the beam to produce *reinforced concrete.* If the steel rods are held under tension while the concrete is setting and later released, the product is called pre-stressed concrete. A pre-stressed concrete beam must be loaded heavily enough to overcome the internal stress from the steel rod before it can begin to bend and crack under the influence of even higher loads. In a similar manner, strong but light composite materials can be built up by compounding innumerable short thin fibres of glass or carbon within a polymer matrix.

compressibility is defined by the following word equation:

$$\text{compressibility} = -\frac{\text{increase in volume}}{\text{increase in pressure} \times \text{original volume}}$$

The SI unit is pascal^{-1}. An increase in pressure compresses the specimen. A decrease in volume is a negative change. The negative sign in front of the definition keeps the compressibility positive. Gases have much larger compressibilities than solids or liquids.

compressive stress is a force per unit area acting on the ends of a material in bar form to push the ends towards one another. It is the opposite of *tensile stress*. It is particularly relevant to materials in the lower parts of a heavy structure which carry the weight of everything above. Bricks used for the tall arches of railway viaducts have to be very strong under compression. They are called engineering bricks.

conductors and insulators have contrasting electrical properties. Conductors are materials in which there are innumerable electrons detached from atoms and forming what is known as an 'electron gas'. Should the conducting material be immersed in an electric field, the electrons will move to new positions within the conductor so as to eliminate the electric field inside the conductor.

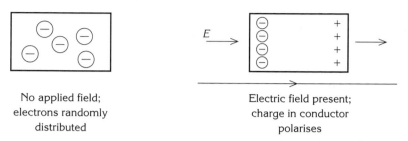

| No applied field; electrons randomly distributed | Electric field present; charge in conductor polarises |

A battery connected across the ends of the conductor has two effects. It sets up an electric field within the conductor which polarises the charge. It removes electrons from the positive end of the conductor and feeds them into the negative end at the same rate. The effect is to sustain the electric field in the conductor by sustaining the flow of electrons. The strength of the electric field depends on the length of the conductor and the potential difference between its ends.

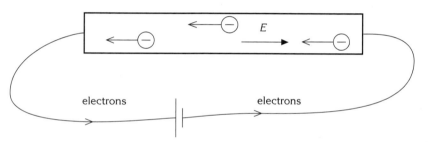

Insulators hold no detached electric charge, no free electrons. The effect of the electric field is to distort the atoms and molecules into electric dipoles which line up with the electric field and reduce its overall magnitude without eliminating it altogether. This polarisation of the insulating material plays an essential part in the action of dielectric materials in capacitors. See *parallel plate capacitor*, *relative permittivity*.

conservation laws (nuclear) are the laws which control the pathway of an event involving subatomic particles. They include conservation of mass-energy, momentum, angular momentum and charge. They also include another set of conservation laws not encountered in the ordinary course of events.

Leptons are not necessarily conserved – they come and go – but lepton number is conserved. Lepton number is conserved by virtue of the antiparticles having negative lepton numbers attached to them. The concept is further complicated by having a separate lepton number for each of the three generations.

Baryon number is conserved but there is a difficulty with mesons. Mesons are built from a quark and an antiquark so they can annihilate. The difficulty is avoided by giving *hadrons* a baryon number 1, antihadrons a baryon number –1 and mesons a baryon number 0.

conservation of energy means that the sum total of energy in the universe is a constant, though it can change its form and flow from one body to another. The first law of thermodynamics is a precise statement of the same principle applied to a finite 'system', not to the universe as a whole.

When the law of conservation of energy is applied to mechanical energy in mechanical systems it is only an approximate law since mechanical energy may be transformed into (say) thermal energy and 'lost' from the system. (See *thermodynamics* (*first law of*).)

conservation of linear momentum is a principle which states that the total *linear momentum* of a system of interacting bodies in any specified direction is constant provided no external force acts on the system in that direction.

This principle is a consequence of *Newton's third law of motion* and always holds true exactly; it is never just an approximation. It seems to be an approximation in 'real life' because it is impossible to create a set of interacting bodies on which there are no external forces. An experiment to illustrate the principle is described in the entry *Newton's third law of motion*.

constant pressure changes are changes at constant pressure. A fixed mass of gas obeys *Charles' law* when undergoing a change of temperature at constant pressure. The volume of the atmosphere is so large, in absolute terms, that most changes in the state of a solid or liquid mass occur at constant pressure. Any change in the state of a gas can be resolved into two changes; one at constant pressure and the other at constant volume. This is one reason why *p–V* diagrams are so useful.

constant volume changes are changes at constant volume. It is rare to have a change at constant volume in the state of a liquid or a solid: the pressures involved are too great. For a fixed mass of gas at constant volume, pressure increases in proportion to the increasing absolute temperature. (See *pressure law*.)

constant volume gas thermometer: a thermometer which uses the principle that the pressure of a fixed mass of gas in a vessel of constant volume is proportional to the *absolute temperature* of the gas. The 'simple' form is the same as the apparatus used to check the *pressure law*. An accurate form has a built-in barometer so that the pressure of the gas is measured directly. It is slow in use, cumbersome but accurate and is used to calibrate other thermometers.

The original attraction of this form of thermometer, which outweighed its large bulk and slow response, were the discoveries that:

- the ideal gas scale of temperature is identical to the absolute thermodynamic temperature scale
- measurements made with a constant volume gas thermometer filled with a real gas such as hydrogen can be corrected to the ideal gas thermometer scale.

This is now history and the constant volume gas thermometer has no more than a modest place in the definition of the *International Practical Temperature Scale*. (See *temperature*.)

constructive interference means that two or more waves meeting at a point enhance one another at all times. For two sound waves to interfere constructively at a point, they must be in phase with one another when they meet at that point and in phase at all times. This can happen only if they are of the same frequency.

Light waves, too, must be in phase where they meet but same frequency will not ensure that this is the case; light waves must also be coherent. The rise and fall of wave displacement for one wave must keep in step with that of the second wave. In practice, this means that the light waves must come from the same source. (See *coherence, destructive interference, interference, principle of superposition*.)

continuous spectrum: a spectrum like that shown in the *total radiation curve* for an incandescent solid, in which radiation of all wavelengths is present. See *absorption spectrum*.

control rods are long, rigid containers filled with a material which has a high absorptive power for neutrons, such as boron. They are lowered into holes drilled in the *moderator* of a nuclear power plant where they control the neutron flux and, as a consequence, the rate of power generation. (See also *thermal fission reactor*.)

convection is a circulatory motion within a fluid mass brought about by temperature gradients. The driving force is the failure of less dense and warmer material lower down to support cooler more dense material higher up. The cool, dense fluid falls pushing the warm fluid upwards. This motion redistributes internal energy. There are no convection currents in weightless circumstances.

The lower layer of the atmosphere (troposphere) is 15 km or so deep and is hotter at the bottom than the top. Convection currents associated with these temperature differences account for much of our wind and weather. The temperature gradient in the next layer up (stratosphere) is the other way round so here there are no convection currents.

coolant is the fluid which circulates around the control rods in a nuclear power plant and which transmits the energy released by the fission processes to the steam boilers of the turbines. See *thermal fission reactor (nuclear reactor)*.

cooling: see *Newton's law of cooling*.

cooling by evaporation is a mechanism by which heat is transferred from a liquid into the surroundings. It is a major factor in transporting solar energy absorbed near the equator towards the poles. There are two key ideas: molecular speeds in a liquid are distributed around an average value and the average value decides the temperature of the liquid. Some of the faster molecules near the surface will always be moving outwards from the liquid, breaking through the surface and escaping. Since only the faster molecules can do this, the average speed of those remaining is diminished, so the temperature of the liquid mass falls.

cooling by working is a mechanism by which a gas can transfer energy to its surroundings. It is a consequence of the *first law of thermodynamics* which can be written and extended as follows:

$$\Delta U = \Delta Q + \Delta W = \Delta Q - p.\Delta V$$

If the gas is expanding, ΔV is positive and ΔW (the work done on the gas) is negative. With ΔQ equal to zero, ΔU will be negative, internal energy falls and the temperature of the gas goes down.

In kinetic theory terms, the gas expands by pushing against the atmosphere. Molecular momentum and kinetic energy with it, are transferred to the atmosphere. The average molecular kinetic energy of the expanding gas diminishes, so its internal energy goes down and so also its temperature.

cosmological principle: the assumption that, at any given instant in time, the universe looks the same in all directions from any position in the universe. This is not to say that the view remains the same with the progress of time; an expanding universe implies a falling density.

coulomb: the SI unit of electric charge. Electric charge is defined by the word equation:

electric charge = current × time

and the coulomb is defined by the same word equation with the units replacing the physical quantities:

coulomb = ampere × second

The symbol for the coulomb is C.

As is always the case with units, *the* coulomb is different from *a* coulomb. *The* coulomb is the unit of electric charge defined above. *A* coulomb is an amount of electric charge – the charge per second flowing past a point in a wire in which a current of one *ampere* is flowing.

coulombmeter: an instrument which measures electric charge in coulombs. In practice the coulombmeter is an application of an op-amp voltage follower circuit. There have been voltage follower circuits almost as long as there have been amplifiers. What makes this circuit special is the enormous input resistance of the op-amp. The leakage current from the input capacitor shown in the circuit below is practically zero.

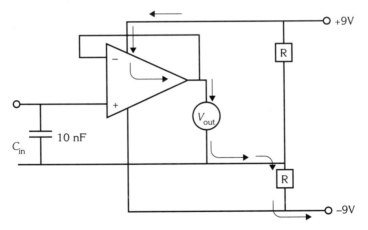

The circuit is often drawn without the power leads on the op-amp. This is a mistake because the action of the circuit is to allow extra current from the power supply to flow through the output (a voltmeter, here) and to increase in value until the output voltage equals the input voltage. For the case when the output voltage stops increasing:

$$V_{out} = V_{in} = Q/C_{in}$$

If C_{in} is known, say 10 nF, and if V_{out} is read from the voltmeter then Q can be calculated.

This, by itself, is not much use. Microcoulombs are not carried around in search of capacitors. The likely situation is that the charge to be measured is already on a capacitor of known *capacitance.* If that capacitor has a much larger capacitance than 10 nF (say 10 µF) and the two capacitors are brought into parallel, a tiny fraction of the charge will leak onto the 10 nF capacitor. The voltage across the 10 nF capacitor equals the output voltage of the op-amp circuit. The charge carried on the larger capacitor can now be calculated. A typical application would be to verify the relationship $Q = CV$.

Coulomb's law states that the force F between two point charges Q_1 and Q_2 a distance r apart in vacuum is given by:

$$F = \frac{1}{4\pi\varepsilon_0} \cdot \frac{Q_1 Q_2}{r^2}$$

where ε_0 is the *permittivity of free space*. The value of ε_0 is defined as $1/\mu_0 c^2$ where μ_0, the permeability of vacuum, is defined to be $4\pi \times 10^{-7}$ H m^{-1}. The value of c, the speed of light, is defined as 299 792 458 m s^{-1} exactly. This sets the value of ε_0 at $8.854\,187\,817 \times 10^{-12}$ F m^{-1}. The importance of these definitions as far as Coulomb's law is concerned is to make all experiment unnecessary except insofar as it would verify the *inverse square law* relationship.

count rate is the number of counts per second recorded by a detector near a radioactive source. It depends on the strength, the distance and direction of the source and on the size and geometry of the detector. Provided neither the source nor the detector are moved, count rate is proportional to activity.

The activity of, for example, an alpha source is the number of disintegrations per second. Many of the alpha particles produced will be absorbed within the source itself. Of those that do leave, only a fraction will be going towards the detector. The detector detects only some of the alpha particles which reach it. This is why count rate is proportional to, but not equal to, the activity.

The unit of count rate is counts per second.

couple is the name given to a pair of equal and opposite forces the lines of action of which are parallel but separated.

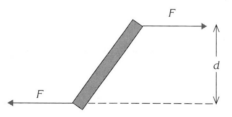

The diagram above shows two equal and opposite forces F acting along parallel lines a perpendicular distance d apart. These two forces form a couple and the moment of the couple is $(F \times d)$ N m. The moment of a couple is a vector quantity (a *pseudo-vector*, in fact, but this does not matter at A level) and the SI unit is the newton metre (N m). This unit should not be confused with the joule; the metre in the definition of the joule runs parallel to the associated force. See *torque*.

covalent bond: a form of attachment between the atoms of a molecule in which neighbouring atoms share electrons. The bond is much stronger than the ionic bond and this strength has two consequences. Molecules of substances such as hydrogen or methane are stable and self-contained: they remain in the gaseous state at low temperatures. The second point is that when a solid is formed containing this kind of bond (as with diamond) the material is exceptionally hard and has a high melting point. (See *ionic bond*, *metallic bond*.)

creep is the slow change of shape of a body under the action of a small but continuous stress acting over a long period of time. An example is the permanent sag that develops slowly along an unsupported metal pipe. The origin of the stress is the pipe's own weight. It happens in glasses, polymers and metals although the atomic mechanisms may differ.

critical angle is the angle of incidence, within a refracting material which marks the onset of *total internal reflection*.

critical temperature is the temperature (different for each gas) above which it is impossible to liquefy the gas no matter how great the compression applied.

cross-linking is the tying together of polymeric chains by atomic bonds along their lengths. Where the binding is by an ionic or covalent bond, it is very strong and makes the polymeric material tough and rigid. This is the case with *thermosetting polymers*. Where cross-linking is primarily via weak *Van der Waals'* forces, the material is generally more flexible and *thermoplastic*. An early example of induced cross-linking is the bonding of *rubber* molecules using sulphur, to make a black rigid material suitable for car tyres or tobacco pipe stems. This is known as vulcanising. The rigidity of battery cases contrasts with the flexible rubber used for car tyres and underlines the importance of controlling the density of cross-linking.

crystalline solids are solids in which an orderly arrangement of atoms extends over so large a volume that the properties of individual crystals decide the properties of the whole. Most salts are ionically bonded solids of this type. These solids have cleavage planes and often display physical properties (thermal expansivity or refractive index) that vary according to direction from which the crystal is viewed. (See *ionic bond, cleaving, crystal defects, slip.*)

Curie temperature is the temperature at which a ferromagnetic material reverts to being a paramagnetic material. The graph below shows (not to scale) how the relative permeability μ_r of a ferromagnetic material such as iron varies with *absolute temperature*. (See *domain theory, relative permeability, paramagnetic substances.*)

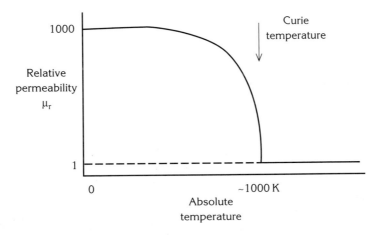

current–time graph: any graph of current against time, but especially a graph showing how the charging current for a capacitor varies with time. The diagram on the next page shows a typical charging circuit for a 500 µF capacitor in series with a 5.0 V supply, a 100 kΩ resistor and a microammeter.

When the circuit is switched on, there is no charge on the capacitor and no potential difference across it. The voltage drop across the resistor is the full 5 V and the current is a maximum (50 µA). As charge builds up on the capacitor so the voltage across it grows. The voltage across the resistor falls and the current flow falls with it.

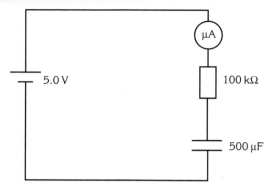

The next diagram shows two graphs: the current–time graph and the charge–time graph for the circuit. The current decays exponentially according to the equation:

$$I = I_{max} \, e^{-t/RC}$$

I drops to I_{max}/e in time $t = RC$. The term RC is called the time constant for the circuit. It tells us that the time taken to charge the capacitor increases in proportion to the series resistance and the capacitance. For this circuit, I_{max} is 50 µA, I_{max}/e is 18.4 µA and the time constant RC is 50 seconds.

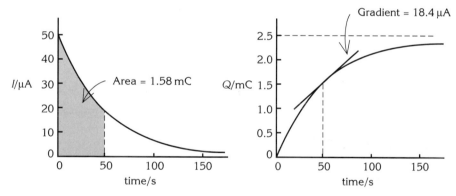

The graph above shows how the charge stored in the capacitor increases towards a saturation value. The charge grows according to the equation:

$$Q = Q_{max} (1 - e^{-t/RC})$$

The labelling on the two graphs underlines the manner in which they are linked. At any time t, the charge amplitude equals the area between the current–time graph and the time axis for the same time period. The current amplitude at the same instant equals the gradient of the charge–time graph at the same instant. The values on the graph are the values at the end of one time constant, 50 seconds.

The charge–time graph approaches a saturation value at a rate which diminishes exponentially with time. The growth of charge with time is not an example of exponential growth.

Exponential growth has a positive power, the gradient increasing with time. (See *time constant*.)

current–voltage characteristics are graphs which show how the current in a circuit component varies with the voltage across it. The three graphs in the diagram below are for a metal wire at constant temperature, a semiconducting diode and a filament lamp. The scales are order-of-magnitude values only.

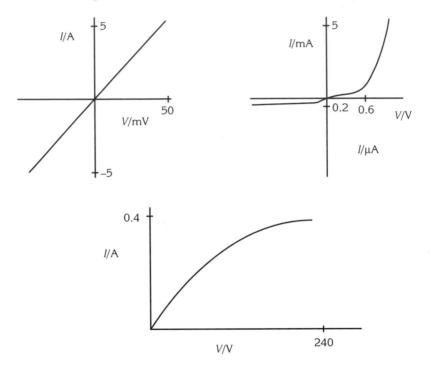

Do you need revision help and advice?

Go to pages 290–306 for a range of revision appendices that include plenty of exam advice and tips.

Dalton's law of partial pressures says that the pressure in a container which holds a mixture of gases in an equilibrium state, equals the sum of the pressures each component gas would exert on its own, in the same container at the same temperature.

If the *kinetic theory of gases* is applied to a mixture of gases, the expression for the pressure becomes:

$$p = \tfrac{1}{3} n_1 m_1 <c_1^2> + \tfrac{1}{3} n_2 m_2 <c_2^2> + \tfrac{1}{3} n_3 m_3 <c_3^2> + \ldots$$
$$= p_1 + p_2 + p_3 + \ldots$$

damped harmonic motion is a *simple harmonic motion* in which amplitude diminishes with time as a consequence of a small damping force impeding the motion. A typical example is a *simple pendulum* oscillating in air. The drag on the pendulum bob is proportional to the speed of the bob so when the bob is stationary, the drag force is zero. Because the speed rises and falls periodically, so does the drag force.

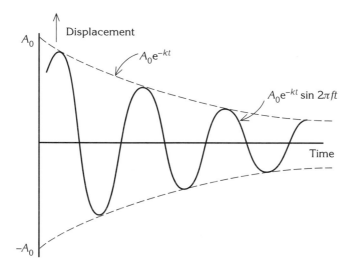

The diagram shows how the displacement of a lightly damped pendulum bob changes with time. The periodic nature of the motion and the independence of period and amplitude are both clear. But when the peak displacements are joined with a smooth curve we see the extra factor introduced by light damping; an exponential decay of the amplitude.

See *exponential change* for a discussion of exponential decay as a widespread phenomenon in physics. Here it is enough to note that the work done per cycle by the pendulum bob against the air resistance is a fixed proportion of the energy stored in the motion at the beginning of the cycle. The energy lost to the air resistance is like a negative interest rate and this explains why the decay is exponential. (See also *resonance, exponential change, forced oscillations*.)

de Broglie wavelength: the wavelength associated with a moving particle and calculated from the relation:

$$\lambda = \frac{h}{mv} = \frac{h}{\sqrt{(2meV)}} \quad \left[= \frac{hc}{eV} \right]$$

where λ is the de Broglie wavelength, h is the Planck constant, m is the particle mass, v is the speed of the particle, e is its charge and V is the potential difference through which it has been accelerated.

The bracketed expression covers the relativistic case. Typical wavelengths for electrons are: 3.9×10^{-11} m at 1 keV, 6.2×10^{-15} m at about 200 MeV and 6.2×10^{-17} m at about 20 GeV. The longer wavelengths are useful for examining crystal and molecular structures. The 200 MeV region is used for studying *nuclear radius* and nucleon diameter. The 20 GeV electrons are used for what is called 'deep inelastic scattering'. These high energy electrons have wavelengths much smaller than the proton diameter and are used to probe inside the proton. Using even higher voltages, quarks are shown to be less than 10^{-18} m in size. See *electron diffraction, particle accelerators*.

decay constant, λ, is the probability per second that a nucleus will decay. It equals $0.693/t_{1/2}$, where $t_{1/2}$ is the *half-life* of the nuclide the nucleus belongs to. See *randomness and radioactive decay, fundamental law of radioactive decay*.

decay rate: a term sometimes used instead of *activity*. It means the same thing.

definition: the meaning of a term or concept stated in a precise verbal or algebraic form, which uses only words or symbols whose meanings are already known.

The equation $F = BIl$, for example, can be used as a definition of magnetic field strength B, because force F, electric current I and length l are all defined and measured independently of this equation. Any formal study of a subject must begin with a few concepts which cannot be defined. The meanings of these concepts are learned by living with them. Straightness, for instance, cannot be defined unambiguously but we all know what the word means. (See *base quantity, derived quantity*.)

density is the mass per unit volume of a material. The symbol for density is ρ and the SI unit for density is kilogram per metre cubed (kg m^{-3}). The word equation definition is:

$$\text{density} = \frac{\text{mass}}{\text{volume}}$$

or, in symbol form, $\quad \rho = \frac{m}{V}$

The densities of most liquids and solids lie within the range 500 kg m^{-3} to 10 000 kg m^{-3}. That is, from about half the density of water to about ten times the density of water. The densities of gases at normal pressure are about a thousand times smaller than the densities of the corresponding liquids.

To measure the density of a material, an object made up of the pure material must be available. If the object is a regular solid, then find its mass and find its volume from its linear dimensions. Calculate the density. For an irregular solid or for a liquid other than water, use a method based on *Archimedes' principle*.

For a gas, weigh an evacuated glass flask and then weigh it again filled with the gas. Find the mass of the gas in kilograms from the difference between the weights in newtons. The volume of the gas can be found by weighing the flask filled with water.

derived quantity: any physical quantity which is not a *base quantity* in SI. Where possible a word equation should be used to define the magnitude of a derived quantity. The same word equation can be used to define the corresponding unit. Capacitance (*C*), and its unit the farad (F) for example, are defined as follows:

$$\text{capacitance} = \frac{\text{charge stored}}{\text{potential difference}}$$

$$\text{farad} = \frac{\text{coulomb}}{\text{volt}}$$

The newly defined quantity, capacitance, is on the left-hand side of the equation. For the definition to be valid, all the terms on the right-hand side of the equation must be previously defined. Their units are known from their definitions. So the units can be written out on the right-hand side of the units equation and the unit on the left-hand side then named. (See *definition*.)

derived unit: a unit in SI for any derived quantity, that is, any SI unit which is not a *base unit*.

The use of word equations to define derived units should ensure that all SI units are products or quotients of SI base units (e.g. m s^{-2} for acceleration). Even so, many commonly used derived units have names of their own. The kilogram metre second^{-2} (kg m s^{-2}) is the unit for force but it is usually called the newton (N). Most such units you need to know are in the following table.

Physical quantity	Name of unit	Symbol	Other units	SI base units
activity	becquerel	Bq		s^{-1}
capacitance	farad	F	C V^{-1}	kg^{-1} m^{-2} s^4 A^2
electric charge	coulomb	C		A s
electric conductance	siemen	S	A V^{-1}	kg^{-1} m^{-2} s^3 A^2
electric potential	volt	V	W A^{-1} J C^{-1}	kg m^2 s^{-3} A^{-1}
electric resistance	ohm	W	V A^{-1}	kg m^2 s^{-3} A^{-2}
energy, work	joule	J	N m	kg m^2 s^{-2}
force	newton	N		kg m s^{-2}
frequency	hertz	Hz		s^{-1}
inductance	henry	H	Wb A^{-1}	kg m^2 s^{-2} A^{-1}
magnetic flux	weber	Wb	V s	kg m^2 s^{-2} A^{-1}
magnetic flux density	tesla	T	Wb m^{-2}	kg s^{-2} A^{-1}
power, radiant flux	watt	W	J s^{-1}	kg m^2 s^{-3}
pressure, stress	pascal	Pa	N m^{-2}	kg m^{-1} s^{-2}

destructive interference means that two or more waves meeting at a point oppose one another at all times and have zero resultant. Two sound waves will interfere destructively at a point where they meet provided that they are out of phase with one another at all times. For this to happen their frequencies must be the same.

For light beams to interfere destructively, two conditions must hold. First, the beams must be coherent if any kind of *interference* is to occur. *Coherence* means that there is a perfect match of phase along the two light beams and, for this to happen, both light beams must come from the same source. Secondly, for the interference to be destructive, the two light beams must be out of phase when they meet. Destructive interference is the mutual cancellation of two beams. It is altogether different from no interference. Two candles give twice the brightness of one wherever their light coincides; the two beams of light neither cancel nor enhance. Light from two different candles is not coherent and will not interfere.

If two light beams are known to be coherent, the *principle of superposition* is employed to find the resultant disturbance at any point where they meet. The principle states how the mathematical functions which represent the two beams are to be added together to find their resultant. (See *constructive interference*.)

diamagnetic substances are weakly magnetised by a surrounding magnetic field but in the opposite sense to the field. The *relative permeability* of a diamagnetic substance is slightly less than unity. A bar of diamagnetic substance will align its long side at right angles to a magnetic field if it is freely suspended. (See *paramagnetic substances*.)

diffraction is the spreading of a beam of waves outside the boundaries set by straight line motion.

The extent to which diffraction occurs depends on the scale of the apparatus compared with the wavelengths involved. A typical Young's two slits arrangement is about a metre long, about ten million times the wavelength of light. The slits themselves are about a thousand wavelengths across. Diffraction effects are observable on this scale, but only with care. The wavelengths of sound range from a few millimetres to a few metres. On a human scale of things diffraction of sound is very marked. The diffraction of sound through a doorway is more like the diffraction of light through a *diffraction grating* slit. The diffraction grating slit is typically thinner than the wavelength of light, diffraction is extreme and the principle of superposition is needed to decide which directions the light will take.

Curiously, diffraction is not an issue with sound largely because its effect is so dominant; sound goes everywhere. With light on the other hand, diffraction effects are barely noticeable but they impose a limit on the power of optical instruments to magnify without loss of detail. In optical instruments, diffraction is a vital consideration. (See *diffraction at a single slit*.)

diffraction at a single slit can be observed with the experimental arrangement shown in the diagram. The slit is illuminated by a parallel beam of monochromatic light (from a laser) and the section of beam which passes through the slit sets up the characteristic intensity pattern shown.

The angular positions θ_n of the successive minima on the screen are given in the equation:

$$d \sin \theta_n = n \lambda$$

where d is the width of the slit, λ is the wavelength of the monochromatic light, and θ_n is the angular displacement of the nth minimum. Between the two first order minima is one broad zeroth order maximum. This central band is twice as wide as all the other bands.

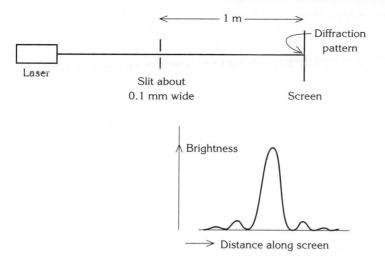

For small angles less than about 5°, sin $\theta = \theta$ provided that θ is measured in radians. Sin 4.5°, for example is 0.07846 and 4.5° converts to 0.07854 radians. So, provided the angles involved are not more than a few degrees, the equation for θ_n becomes $\theta_n = n\lambda/d$ and the minima will be equally spread about the principal axis.

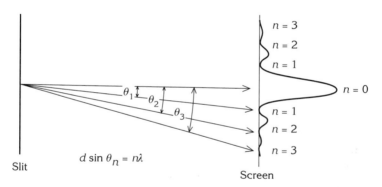

Although the minima are evenly spread in terms of angular displacement, the amplitudes of the illuminated bands between them fall away very fast. You will need a dark room and good equipment to see more than two or three bands on each side of the maximum.

diffraction at a small circular hole in an opaque sheet generates, on a screen, a pattern of bright and dark rings about a bright spot, as shown in the diagram.

The radius of the nth dark ring and radius of the first dark ring are given by the following equations:

$$\sin\theta_n = \frac{122n\lambda}{d} = \frac{r_n}{d} \qquad \text{and} \qquad \sin\theta_1 = \frac{122\lambda}{d} = \frac{r_1}{d}$$

where d is the diameter of the hole in the screen, D is the distance of the screen from the hole, λ is the wavelength of the light used, r_n is the radius of the nth dark ring and θ_n is the angle subtended at the centre of the hole by r_n. The r_n/D and r_1/D factors are correct for small angles only.

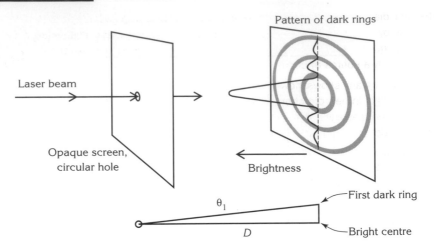

Pattern of dark rings

Laser beam

Opaque screen, circular hole

Brightness

θ_1

D

First dark ring

Bright centre

Imagine what happens as the hole gets bigger. The diameter d increases and θ_n diminishes. Eventually, the diffraction pattern is swamped by the light flooding through the large hole. This is no good. But if there is a lens in the hole, focusing all the light onto one spot, we are back to seeing the diffraction pattern again. This is what happens in the *eye* and in the *telescope*. In these cases the distance D is the focal length of the lens involved and d is its diameter (d is the pupil diameter for the eye and the objective lens diameter for the optical telescope). For a radio telescope, the wavelength λ is very large and d has to be very large too; it is the diameter of the radio telescope dish.

A similar diffraction pattern of light and dark rings is seen if a laser beam is shone through a glass sheet over which opaque particles of roughly equal size have been scattered. Lycopodium powder is often used. If the wavelength is known and θ_1 is measured, then d can be calculated. In other words, the preferential scattering of light by diffraction into particular angles gives information about particle size. See *electron diffraction*.

diffraction grating: a sheet of glass, about 4 cm square, with thousands of equidistant parallel lines etched on one side. Typically, the distance between the centres of two adjacent lines is two or three times the *wavelength* of light.

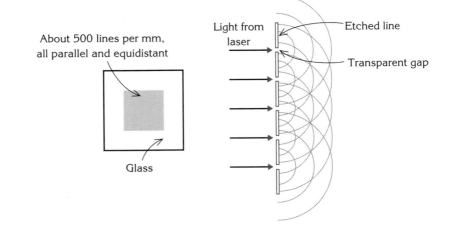

About 500 lines per mm, all parallel and equidistant

Glass

Light from laser

Etched line

Transparent gap

The first diagram shows a plan and magnified section of a diffraction grating illuminated normally by a parallel beam of monochromatic light from a laser. The etching is shown as solid (opaque) lines and the transparent parts as gaps. The gaps are narrow and each one is effectively a source of circular waves.

The next diagram shows how these circular waves set up new wavefronts and send out beams of light in new directions. The first diagram is the so-called zeroth order beam or the zeroth order spectrum. The next diagram shows the formation of the first order spectrum with the relevant equation. There is then the second order and next the nth order.

In general, if λ is the wavelength of the light transmitted by the grating, if d is the distance between the centres of adjacent slits and θ_n is the angle between the nth order beam and the zeroth order beam, then:

$$d \sin \theta_n = n\lambda$$

Alternatively, since $d = 1/N$, where N is the number of lines per metre of grating,

$$\sin \theta_n = Nn\lambda$$

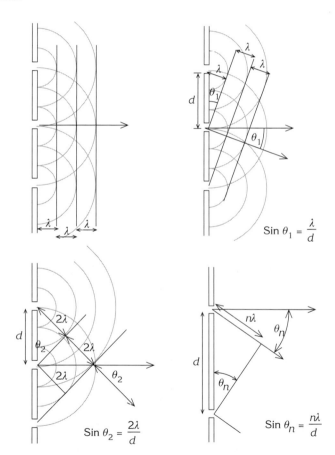

$$\sin \theta_1 = \frac{\lambda}{d}$$

$$\sin \theta_2 = \frac{2\lambda}{d}$$

$$\sin \theta_n = \frac{n\lambda}{d}$$

These equations can be applied either to find the angles of the several orders of spectrum for a monochromatic beam or to find the dispersion in the first order beam, say, for white light.

Worked examples:

1. A diffraction grating has 400 000 lines per metre and it is illuminated normally by light of wavelength 600 nm from a laser. Calculated the number of orders of spectrum formed and the angles to the principal axis at which they leave the grating.

    Find n for θ_n equal to $90°$.

    Since sin 90 equals 1, $n = 1/N\lambda = 1/[(400\ 000\ \text{m}^{-1}) \times (600 \times 10^{-9}\ \text{m})]$.

    This gives n equal to $1/0.24$ or 4.2 to two significant figures. Since n must be an integer, this means that there will be a spectrum for $n = 4$ but not for $n = 5$. If the zeroth order is included, five orders of spectrum are formed.

    The angle of the first order spectrum is found as follows:

    $$\sin \theta_1 = Nn\lambda = (400\ 000\ \text{m}^{-1}) \times 1 \times (600 \times 10^{-9}\ \text{m}) = 0.24$$
    $$\theta_1 = \sin^{-1} 0.24 = 13.9°$$

    The other angles are found in a similar manner:

    $$\theta_2 = \sin^{-1} 0.48 = 20°$$
    $$\theta_3 = \sin^{-1} 0.72 = 46°$$
    $$\theta_3 = \sin^{-1} 0.96 = 74° \text{ (to 2 s.f.)}$$

2. The wavelengths of visible light range from 400 nm for deep violet to 700 nm for the far red. Find the dispersion in the first order spectrum of white light formed by a diffraction grating with 500 000 lines per metre.

    Following the procedure used in the last example:

    $$\theta_{\text{deep violet}} = \sin^{-1} 0.200 = 11.5°$$
    $$\theta_{\text{far red}} = \sin^{-1} 0.350 = 20.5° \text{ (to 3 s.f.)}$$

    and the dispersion is $9.0°$.

digital signal: a signal that is made up of bits. Each bit has two possible values labelled 1 or 0, HIGH or LOW, ON or OFF, ACTIVE or INACTIVE. Examples of a one-bit digital signal are the red indicator lights on an electric iron and on a car dashboard. Flashing car direction indicators are a one-bit signal, either ACTIVE or INACTIVE. Two such one-bit signals make a two-bit signal which can be coded to identify up to four different states (e.g. straight ahead, turn left, turn right, watch it). Binary numbers are used to identify the different possible states. A two-bit signal might be 00, 01, 10, or 11. A four-bit signal (called a nibble) can label up to 16 states and an eight-bit signal (called a byte) can label up to 256 possible states (i.e. 0 to 255 put into binary numbers).

In digital electronics, a byte usually takes the form of eight square voltage pulses, transmitted in sequence. Each pulse has one of two possible magnitudes and each pulse lasts for the same time interval. If the magnitudes are labelled 1 or 0, then the digital signal in the diagram reads 10110011 (i.e. 128 + 32 + 16 + 2 + 1 = 179).

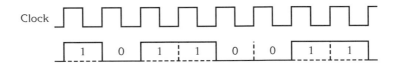

The timing of a digital signal and its length are governed by a 'clock'; the clock is a square wave whose rises and falls in voltage are governed by crystal oscillator. A clock signal generated by a quartz crystal is what keeps electronic watches so accurate. The clock signal is a vital component of any digital circuit. It controls the timing of a digital signal, its progress within a circuit and its relationship with other digital signals in the circuit. The close control is achieved by using either the rising edges or the falling edges of the clock signal to activate digital switches.

A digital signal that carries information is usually the output signal from an *analogue-to-digital converter*. The digital output is a coded record of the amplitude of the analogue voltage input. Encoding an analogue signal in digital form has many advantages: a digital signal

- can be cleaned of any noise it is exposed to during transmission

- can be stored in a computer memory

- can take part in a computer controlled process and

- can be kept in step with a clock in a digital circuit.

digital-to-analogue conversion is the reverse of *analogue-to-digital conversion*. It is sometimes referred to as 'decoding'. The output may suffer from the information loss associated with the *quantisation* effect at the analogue-to-digital conversion stage.

digital transmission systems transmit a voltage signal in the form of a succession of square pulses. Each pulses has one of two possible amplitudes, often labelled HIGH or LOW. The usual process is for eight successive bits to represent in binary form the magnitude of an analogue amplitude at the moment of sampling. It follows that the bit transmission frequency must be much higher than the sampling frequency unless there is some species of multiplexing. The principal advantage of digital transmission systems is the efficiency with which noise can be removed from the transmitted signals. One disadvantage is the extra expense of the additional electronic activity, another is the poor resolution in a *digital signal* for small changes of amplitude leading to a larger relative error at low amplitudes. (See *time-division multiplexing*.)

diode: an electrical component which conducts current in one direction but not the other. Early diodes consisted of two electrodes (hence the word diode) fitted into a glass bulb which was evacuated or, later, filled with inert gas. One electrode, the cathode, was heated to dull red heat and thereby empowered to release electrons by *thermionic emission*. The second electrode, the anode, was not heated. A small positive potential difference would enable electrons to flow from the cathode to the anode but not the other way round.

Most modern diodes are solid state devices built up from semiconducting materials. See *semiconducting diode, light-emitting diode*.

dislocations are interruptions to the ordered structure of a crystalline solid. They may occur at a point (point defects), extend along a line (edge dislocations) or spread across an area (grain boundary).

Point defects can be associated with:

- a space where an atom ought to be (vacancy)

- an impurity atom replacing an ordinary atom in a crystal (substitution)

- an impurity atom trapped between two crystal planes (interstitial).

Whatever the source of the trouble, there are strains set up in the neighbouring parts of the crystal and the point dislocation is a source of weakness.

Edge dislocations are found within a crystal where one plane suddenly ends and the planes on either side bend over the edge to close up the space left. This is a source of considerable weakness in a single crystal. Most metals are assemblies of thousands of tiny crystal called grains and whose surfaces are called grain boundaries. Because dislocations cannot cross grain boundaries, a tight grain structure inhibits the flow of edge dislocations in the presence of stress and adds to the strength of the material as a whole.

Grain boundaries are the surfaces of small single crystals which make up the structure of a metallic solid. In practice, grains prevent the flow of the material when subjected to stress and the smaller the average grain size the stronger the material becomes. This is because the other forms of dislocation become ineffective.

dispersion of light is the separation of a beam of light into its component wavelengths or colours. This is possible because the speed of light in a substance such as glass changes slightly from one wavelength to the next. The dispersion of white light by glass is clearly demonstrated by the formation of a spectrum using a glass prism. This is shown in the diagram below.

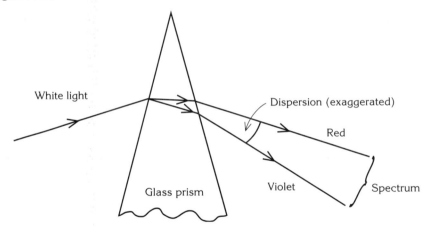

(See *electromagnetic spectrum*.)

dispersion of sound means the change of speed with frequency. It is zero for audible frequencies (speech and music would both fail over modest distances were this not the case) but it is measurable at much higher frequencies. There is some slight dispersion of sound in humid air, this is linked to the absorptive action of the water vapour present. (See *sound waves*.)

displacement defines the position of one point with reference to another point. It is a *vector quantity*. Its magnitude is the straight line distance between the two points. Its direction is the angle between the straight line joining the two points and a reference direction. The SI unit for displacement is the metre, symbol m.

The second part of the diagram underlines the difference between 'distance travelled' and 'change of displacement'. Say a dog walks from a point A to a point B along a curved path of length d. The distance travelled is d metres. But the change of displacement, Δs metres, is the vector difference between the initial displacement s_A and the final displacement s_B.

The change of displacement depends only on the start and finish points. It is independent of the path followed by the dog. It is a vector difference:

$$\Delta s = s_B - s_A$$

Had the dog returned to its starting position, the points A and B would coincide and the change of displacement would have been zero. The average velocity would have been zero. But the average speed could be anything from very slow to as fast as a dog can run.

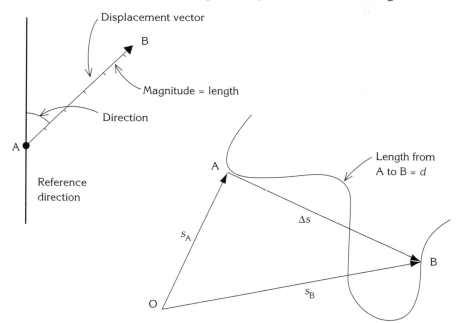

displacement–time graph: a graph which shows how the *displacement* of an object varies with time. The displacement axis of a displacement–time graph gives the magnitudes for displacement but says nothing about direction, other than forwards or backwards, up or down. Consequently, these graphs describe motion in one direction only. They can be used for either *straight line motion* or one component of a more complex line of travel.

For straight line motion, the gradient of the tangent to a displacement–time graph is the *velocity* at that instant corresponding to the point at which the tangent is struck. For other than straight line motion, the gradient of the tangent is the component of the velocity in the same direction as the displacement. If displacement decreases with time, the velocity is negative (see the examples on page 62).

The diagram on page 62 shows a displacement–time graph and the corresponding *velocity–time* and acceleration–time graphs, for a glider on a horizontal linear air-track bouncing backwards and forwards elastically between reflectors which are 1.20 m apart. Notice that a small adjustment for the zero speed at the moment of bounce has to be included. The displacement increases and decreases uniformly, the velocity is constant but changes direction and the acceleration is zero except at the moment of bounce. The details of the bouncing action are not exact.

The steady velocity is slightly higher than 0.8 m s^{-1} or slightly lower than –0.8 m s^{-1}. The gradients of the straight parts of the displacement–time curve equal the steady velocity values and the area of each loop in the velocity graph will equal 1.2 m. The direction of acceleration is known, but its peak magnitude is not; this is why there is no scale on the acceleration graph.

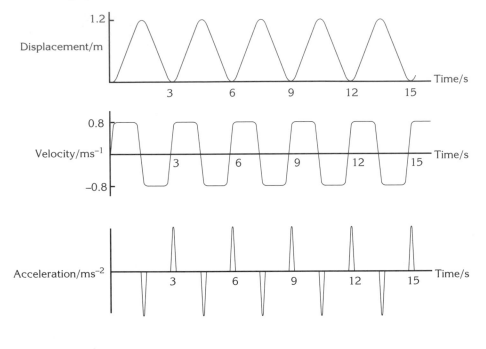

Worked example

A student inclines a linear air-track at a small angle to the horizontal and releases a glider 1.8 m from an elastic reflector at the lower end. The times at which the glider passes particular points on the track are noted from the starting time until the glider returns to its starting position. A displacement–time graph is drawn up and used to draw velocity–time and acceleration–time graphs.

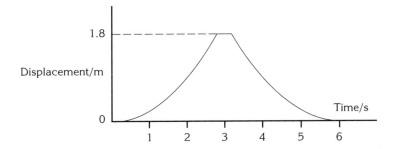

Notice that the final displacement is zero. The velocity suddenly changes direction at the reflection but a slight slowing down as it returns uphill is equivalent to an increasing downwards speed. The acceleration is constant throughout the motion except during the bounce when it goes through two negative peaks separated by a zero period. (The value of the bounce time interval was set to keep the time axis tidy.)

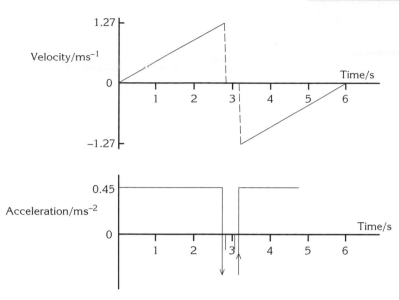

distance tells us how far along a line one point is from another. The line may be straight or curved. Distance is a *scalar* quantity. The SI unit for distance is the metre, symbol m.

Mechanics is mostly concerned with either *distance travelled*, another scalar quantity, or with *displacement*, a vector quantity. Distance is involved more with topics such as the inverse square law, attenuation and diffraction patterns.

distance–time graph: a graph which shows how distance travelled along a line varies with time (it should be called a distance travelled – time graph, but never is). The gradient of the tangent to a distance–time graph at an instant is the speed of the object at that instant. The diagram below illustrates three important cases, zero speed, constant speed and constant rate of increase of speed.

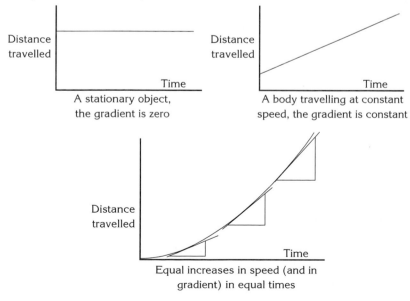

A stationary object, the gradient is zero

A body travelling at constant speed, the gradient is constant

Equal increases in speed (and in gradient) in equal times

Remember that the s in the equations of motion stands for *displacement* and that, mostly, you will be concerned with *displacement–time graphs*. Distance travelled can only increase or stay constant; displacement can increase or decrease and even go negative.

distance travelled is not simply a *distance*. Walk four paces forwards, two backwards and six forwards. The distance from start to finish is eight paces. But the distance travelled is twelve paces.

Distance travelled is measured along the travel path. Only the magnitude is recorded. The line may be curved or straight and the direction is irrelevant. The SI unit of distance travelled is the metre. It is a scalar quantity.

Speed is defined as distance travelled per unit time. Since distance travelled can only increase, speed must always be positive. (See also *displacement*.)

distribution of speeds: a statement of the probable number of molecules, n_v δv, in a fixed mass of gas at a fixed temperature which have speeds within the range v to $v + \delta v$. The information is frequently summarised in a graph

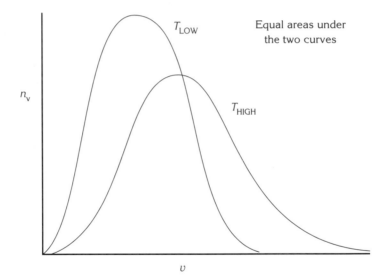

The graph above shows two distributions for the same sample of gas, one at low temperature and one at high temperature. The area between the curve and the speed axis equals the number of molecules in the sample. Two curves for the same sample of gas at two different temperatures must have equal areas. The curve for the higher temperature has more high-speed molecules; it is wider but lower.

(The naming of the ordinate for this graph is a bit tricky. The number of molecules with speed exactly equal to a particular value v, is probably zero and certainly unknowable. What can be calculated is the probability that the speed of a molecules will lie between v and $v + \delta v$. Divide this number by δv and multiply it by N, the number of molecules in the sample and you have n_v. n_v is the number of molecules per unit velocity interval with speed v.)

The mathematical equation for the distribution of speeds is called the Maxwell–Boltzmann distribution law. Using this law it is possible to derive the relationship between the root *mean square speed* c_{rms}, the most probable speed c_0 and the average speed c_{av}. These are:

$$c_{av} = c_{rms} \times \sqrt{\left(\frac{8}{3\pi}\right)} = 0.92\, c_{rms} \quad \text{(to 2 s.f.)}$$

$$c_0 = c_{rms} \times \sqrt{\left(\frac{2}{3}\right)} = 0.82\, c_{rms} \quad \text{(to 2 s.f.)}$$

domain theory accounts for the magnetic properties of ferromagnetic materials. It is an ancient theory and began with the proposition that materials such as iron are constructed from elementary magnets. When these tiny magnets are aligned in the same direction, the specimen is magnetised and shows its magnetism on a human scale. When the elementary magnets form closed magnetic rings within the specimen, their combined effect outside the specimen is reduced to nothing.

Modern science has added little to this so-called 'molecular theory of magnetism' except to identify the elementary magnets as small molecular structures known as domains. The domains can be revealed by etching and observed when highly magnified.

Should a ferromagnetic material be heated, there comes a temperature at which the domains are broken down by thermal motion. The material loses its ferromagnetic properties and becomes *paramagnetic*. This temperature is called the *Curie temperature*. The Curie temperature for iron is about 770°C.

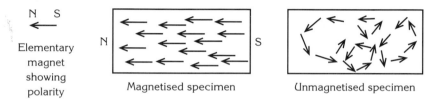

Elementary magnet showing polarity

Magnetised specimen

Unmagnetised specimen

Doppler effect: the change in *frequency* (and of *wavelength*) of any wave motion when the source of the waves or the observer, or both, are in motion. With sound, the immediate observation is a change of pitch. The relevant equation for sound in still air is

$$\frac{f_o}{f_s} = \frac{v \pm v_o}{v \mp v_s}$$

where f_o is the observed frequency, f_s is the source frequency (i.e. the frequency of oscillation of the loudspeaker or whatever), v is speed of sound in still air and v_o and v_s are the speeds of the observer and the source relative to the air. The upper signs on the right hand side are for motion towards, the lower signs are for motions apart. The relationship is more complicated if there is an attempt to allow for wind speed.

With light or any other electromagnetic wave, it is usual to measure wavelength rather than frequency. The Doppler effect (called a Doppler shift) is expressed as a wavelength change. The relevant equation is:

$$\frac{\Delta\lambda}{\lambda} = \pm\frac{v}{c}$$

If the source is moving towards the observer, the − sign is used. The wavelength shortens and the colour shift would be towards the blue. If the source is moving away from the observer, the + sign is used. The wavelength increases and the colour shift would be towards the red.

The main application of the Doppler effect in *astronomy* is to measure line-of sight veloci-ties. Line-of-sight velocity added vectorially to sideways velocity gives actual velocity, allowing relative motions within our galaxy can be examined. On the larger scale, the red shift of light from other galaxies indicates that line-of-sight velocities of other galaxies are always away from the galaxy that contains the solar system. See *Hubble's law*.

drag is the frictional force which opposes the motion of an object through a fluid. If the object is a sphere and its speed equals the terminal speed, provided that the flow is streamline then the magnitude of the drag is given by *Stokes' law*. Drag is dramatically increased by the onset of turbulence and considerable attention is paid to this matter in vehicle design. (See also *aeroplanes*, *turbulence*.)

drift speed is the average speed of progress of charge carriers through a conductor when current is flowing in the conductor. If I is the current in a conductor of uniform cross-section A, having carrier density n with each charge carrier having charge q and moving with drift speed v, then:

$$I = nAvq$$

Put I equal to 5 A, n at 1×10^{29}, A at 1×10^{-6} m² and q at 1.6×10^{-19} C and v works out at about 0.3 mm s^{-1}. Typical drift speed values in conductors range from about 0.01 mm s^{-1} up to about 100 mm s^{-1}. They are much higher in semiconductors because the carrier density n is much smaller.

ductility is the capacity of a metal especially, and of other substances occasionally, to be drawn out into a thin wire.

Do you need help with the synoptic element of your Physics course?

Go to page 296 for tips and advice.

ear: the organ of hearing. The functions of the three main parts of the human ear are collection (the outer ear), amplification (the middle ear) and detection (the inner ear).

Sound is almost wholly reflected by the human body. This is because the acoustic impedance of the body is high and that of the surrounding air is low. The remedy in the ear is to have a membrane of low acoustic impedance mounted across a pathway for the sound. The outer ear collects sound across an area and funnels it into the ear tube towards the tympanic membrane. The ear tube has a resonant frequency close to 1 kHz and this strongly influences the frequency response of the ear. The graph in the diagram on page 68 shows how the ear responds to different frequencies from low to high. Although the human voice is carried on frequencies in the low hundreds, all the detail is carried in the 1 to 10 kHz region and the ear must respond accurately to this frequency range.

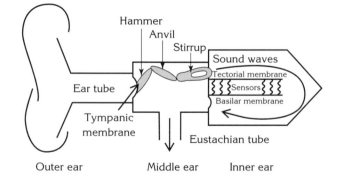

The middle ear amplifies the sound in the sense of transforming a movement of large amplitude and low pressure at the tympanic membrane into a movement of small amplitude but high pressure at the entrance to the inner ear, the oval window. The necessity for this process lies in the high acoustic impedance of the inner ear. Amplification is effected by a lever system constructed from three bones, the hammer, the anvil and the stirrup. The pressure within the middle ear is kept equal to the pressure outside by the Eustachian tube which opens into the back of the throat. This precaution prevents the possible build up of an appreciable pressure difference across the delicate membrane.

The inner ear is divided into two pathways by the basilar membrane. The nerve structures which sense the sound stretch between the basilar membrane and the tectorial membrane with low frequencies detected early and high frequencies detected late.

The ear reports two sensations to the brain, pitch and loudness. Pitch is a response to the dominant frequency in the sound. Loudness is a response to the intensity (i.e. the sound energy carried into the ear per square metre per second). The minimum intensity to which the ear can respond varies, but the notional minimum level I_0 is defined as 10^{-12} W m^{-2}. The loudness L in decibels of a sound of intensity I is defined by the formula:

$$L = 10 \log_{10} \frac{I}{I_0}$$

L is around 50 dB in ordinary conversation, 80 dB is getting too noisy and 100 dB is about the level of loud thunder. Around 120 dB is painful and 140 dB causes irreversible damage. Anything above 90 dB is damaging in the long term (loud music, pneumatic drills etc.)

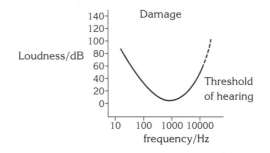

If you study biology, reflect on the differences between fish and ourselves as far as hearing and breathing are concerned. The acoustic impedance of a fish is much the same as that of the surrounding water. Any vibrations carried through the water pass into the fish's body and are easily sensed. So fish don't need ears. But they need gills to get oxygen out of the water. Evolution has provided us with lungs to breathe: we don't need gills. But we do need ears. So gills have evolved into ears. The passageway from the back of the fish's throat to the gills has evolved to become the Eustachian tube. The cartilage in the gill flaps has evolved into the three bones of the middle ear.

eddy currents are induced currents in metallic supporting parts of electrical machinery such as motors or transformers. These metal parts either rotate at speed in a magnetic field or stand in a rapidly changing magnetic field. The induced e.m.f.s set up in the conductor drive current loops known as eddy currents through the metal. These current loops will, unless they are diminished by good design, represent a serious energy loss. In the case of transformer cores, eddy current loss is reduced by constructing the cores from thin steel sheets insulated from one another.

efficiency (machines): the ratio of the useful work done by a *machine* (such as a pulley system) on a load (such as a weight) to the work done on the machine by the effort force. It is usually expressed as a percentage.

$$\text{efficiency} = \frac{\text{work done on load}}{\text{work done by effort force}} \times 100\%$$

If a machine has an efficiency of 80%, the implication is that about a fifth of any effort applied to the system is expended working against friction.

efficiency (thermal power station) is the ratio of the electrical energy reaching the consumer from the thermal power station to the internal energy released when the primary fuel burns. The main losses are:

- energy released in the furnaces which is not carried to the turbines by the steam

- inefficiency in the conversion of energy carried in the steam into electrical power

- power losses during transmission of the electrical power.

Einstein's photoelectric equation: see *photoelectric effect.*

elastic collision: a *collision* in which *kinetic energy* is conserved. If a collision obeys a different constraint, such as the two bodies sticking together on contact, then the collision cannot be elastic. The law of *conservation of linear momentum* is always obeyed.

elastic limit: the maximum stress which can be applied to an elastic specimen without causing a permanent change in the dimensions of the specimen. (See *stress–strain curve.*)

elastic strain energy is the energy stored in a material under stress. If the stress is tensile in nature then the energy stored equals ½ *F x* where *F* is the tensile force applied to the specimen and *x* is the extension of the specimen. (See *force–extension curve.*)

elasticity is the ability of a material to recover its original shape or size once a deforming stress has been removed. (See *stress–strain curves, plasticity.*)

electric charge is a fundamental property of matter. Amount of electric charge is defined by the word equation:

electric charge = electric current × time

The ampere and the second are *base units*. The same word equation defines the SI unit of electric charge, the coulomb:

coulomb = ampere × second

(See *Coulomb's law, electric field.*)

electric charge and electric current are related physical quantities; electric current is the rate of flow of electric charge. Initially electric charge was studied in the static condition, distributed across the surface of a charged body. When different objects were charged by friction it was soon realised that there were two kinds of charge, now labelled positive and negative, and that like charges repel while unlike charges attract. At a later stage it was discovered that matter itself is electrical; atoms consist of negative electrons in their outer regions shielding positive nuclei at their centres.

An analysis of the forces between charged bodies led to the concepts of *electric field*, *electric field strength* and *electric potential*. Experiments with charged bodies led to the distinction between *conductors and insulators* and to the concept of *capacitance*.

Later still, simple electric cells were discovered. Experiments involving the rapid and repetitive charging and discharging of conductors established that electric current is flow of electric charge. The discovery of the *first law of thermodynamics* and energy conservation led to the definition of the volt, a stronger interpretation of electric potential and to the analysis of energy transformations in electric circuits. The final steps were the discovery of the *electromagnetic spectrum* and then the laws of relativity.

Electric current equals charge transferred per second. This idea is summarised mathematically in the equation:

$$I = \frac{\Delta Q}{\Delta t} \quad \text{or} \quad \Delta Q = I \Delta t$$

and, for a current in a uniform wire, this equation reduces to:

$$I = nA\upsilon q$$

where I is the current, n is the number of charge carriers per unit volume of wire, A is the area of cross-section of the wire, v is the average speed of the charge carriers (called *drift speed*) and q is the charge on the carriers.

For a beam of electrons, this equation becomes:

$$I = ne\upsilon$$

where e is the electronic charge and n is now the number of electrons per unit length of beam. Both equations follow immediately from the definition of current as the rate of flow of charge.

electric circuit: a complete loop of conducting material which includes a current source and a current detector. In his first historic experiments, Galvani brought together by accident a simple battery (a moist iron nail piercing a copper sheet), a completed circuit (a frog's leg in contact with the copper sheet and the iron nail) and a detector (the nervous twitching action of the frog's leg).

Complex electric circuits are analysed algebraically by applying *Kirchhoff's first law* to the junctions in the circuit and *Kirchhoff's second law* to the smallest number of circuit loops, which, taken together, include every component in the complete circuit.

For simple circuits, it is enough to be familiar with two equations, *Ohm's law* and the power equation. Both equations can be applied either to individual components or to the circuit as a whole. These equations for a single component are written:

$$V = IR$$

volts = amperes × ohms

$$P = IV = I^2R = V^2/R$$

watts = volts × amperes

where V is the *potential difference* across a component of resistance R, I is the current flowing in the component and P is the power dissipated in the component. Although it is general practice to assume that Ohm's law applies unless we are told otherwise, its strict application assumes that the temperature of the resistor is somehow held constant. Ohm's law applied to a complete circuit made up of a current source and resistors leads to a longer statement of Ohm's law. The circuit e.m.f. E equals the sum of the potential differences V taken round the circuit:

$$E = V_1 + V_2 + V_3 + \ldots$$
$$= I_1R_1 + I_2R_2 + I_3R_3 + \ldots$$

This equation is fundamentally an energy statement. E is the energy in joules provided by the power supply to each coulomb of charge that loops the circuit. This energy is dissipated in the circuit components, including the power supply itself; and each potential difference V across a component is the energy per coulomb and in joules dissipated in that component. Any energy dissipation within the voltage source is looked after by the concept of *internal resistance*. Internal resistance accounts for any difference between e.m.f. E and terminal potential difference V.

If R represents the equivalent resistance of a complete circuit, we may write:

$$E = IR$$

Then, multiplying both sides by current I, we have:

$$EI = I^2R = P$$

It is helpful to read this equation through as a word equation:

(energy per coulomb) $\times$ (coulombs per second) = (energy per second) = power P

The longer form of the equation for a series circuit will be:

$$EI = V_1I + V_2I + V_3I + \ldots = P$$

or $\quad P = EI = I^2R_1 + I^2R_2 + I^2R_3 + \ldots$

(See *ampere, coulomb, e.m.f., volt.*)

electric field: a region in which a force would act on an electric charge. The strength E of an electric field is defined by the word equation:

$$\text{electric field strength} = \frac{\text{force}}{\text{charge}}$$

$$E = \frac{F}{Q}$$

This definition of electric field strength also defines its SI unit as the newton per coulomb (N C^{-1}). An alternative SI unit for electric field strength is volt per metre (Vm^{-1}). Since *electric field strength* and force are both vectors, it is assumed that the charge involved is positive. Also, the charge must be of infinitesimal size if it is not to distort the electric field whose strength it is being used to measure. The formal definition becomes:

'The electric field strength at a point is given, in magnitude and direction, by the force per unit charge acting on a very small positive charge placed at the point.'

The force F applied to a charge Q_2 by a charge Q_1 a distance r away is given by Coulomb's law:

$$F = \frac{1}{4\pi\varepsilon_0} \cdot \frac{Q_1 Q_2}{r^2}$$

But

$$F = EQ_2$$

The electric field strength E a distance r from a positive charge Q is evidently given by the equation:

$$E = \frac{1}{4\pi\varepsilon_0} \cdot \frac{Q}{r^2}$$

and this equation leads directly (though the derivation is omitted) to an expression for the electric potential V a distance r from a positive charge Q:

$$V = \frac{1}{4\pi\varepsilon_0} \cdot \frac{Q}{r}$$

The electric potential V at a point a distance r from a positive charge Q, is the work done per unit charge bringing a very small positive charge from infinite distance to the point. This definition is a special case of the general definition for electric potential difference:

'The electric potential difference **V** between two points equals the work done per unit charge bringing a very small positive charge from one point to the other.'

We can now write down a word equation for potential difference:

$$\text{potential difference} = \frac{\text{work done}}{\text{charge}}$$

The same equation defines the unit for potential difference, the volt:

$$\text{volt} = \frac{\text{joule}}{\text{coulomb}}$$

electric field strength, E, at a point is given, in magnitude and direction, by the force per unit charge acting on a very small positive charge placed at the point. The word equation definition is:

$$\text{electric field strength} = \frac{\text{force}}{\text{charge}}$$

$$E = \frac{F}{Q}$$

The SI unit of electric field strength is newton per coulomb (N C^{-1}) or volt per metre (V m^{-1}). Electric field strength near an isolated positive point charge Q follows an inverse square law and is as follows:

$$E = \frac{1}{4\pi\varepsilon_0} \cdot \frac{Q}{r^2}$$

(See *electric field*.)

electric motor: a machine which transforms electrical energy into rotational mechanical energy by exploiting the force on a current-carrying conductor in a magnetic field.

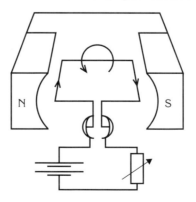

The diagram displays in simplified form the principle of the DC electric motor. The current is fed to the coil through a split ring commutator which ensures that the left-hand half of the coil always receives current and the right-hand half of the coil always returns it. A soft iron cylinder (not shown) between the curved poles of the magnet creates a radial field. This ensures that the plane of the coil is parallel to the magnetic field direction except when its inertia carries it through the vertical position. An application of *Fleming's left hand rule* to this situation shows that the force on the left-half of the coil is always downwards and that the force on the right-hand half of the coil is always upwards. The coil is forced by the current and the magnetic field to rotate in an anti-clockwise sense and this can be used to drive machinery and do work.

(See also *generator*.)

electric potential, V, at a point a distance r from a positive charge Q is the work done per unit charge bringing a very small positive charge from infinite distance to the point.

A little mathematics shows that:

$$V = \frac{1}{4\pi\varepsilon_0}\cdot\frac{Q}{r}$$

If r is infinitely large, V goes to zero.

Electric potential is a special case of electric potential difference, which is defined as follows:

'The electric potential difference **V** between two points equals the work done per unit charge bringing a very small positive charge from one point to the other.'

We can now write down a word equation for potential difference:

$$\text{potential difference} = \frac{\text{work done}}{\text{charge}}$$

The same equation defines the unit for potential difference, the volt:

$$\text{volt} = \frac{\text{joule}}{\text{coulomb}}$$

The force F on positive charge Q at a point where the electric field strength is E is given by:

$$F = EQ$$

and the work done moving this charge a distance dx against the direction of the electric field strength E is:

$$dW = Fdx = -EQ\,dx$$

The negative sign appears because E and dx are in opposite directions.

But dW = Q dV, where dV is the difference in potential between the two points. (This is how dV is defined.) These two expressions for dW can be equated to give:

$$QdV = -EQ\,dx$$

or,

$$E = -\frac{dV}{dx}$$

This important result states that electric field strength equals the potential gradient at a point and explains why a possible SI unit for electric field strength is volt per metre (V m^{-1}).

electrical circuit calculations

This analysis can be extended to give an expression for the rate of energy dissipation when current flows in a resistor. The diagram below shows a conductor of resistance R carrying current I in response to a potential difference V.

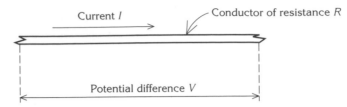

The energy expended by the current source in driving charge Q through resistance R across the potential difference V is W, where:

$$W = QV$$

But

$$Q = It$$

where t is the time for which the current flows. We now have:

$$W = IVt = I^2Rt$$

and the rate of energy expenditure, the power P, equals

$$P = IV = I^2R$$

In terms of units we have:

$$\text{joules} = \text{amperes} \times \text{volts} \times \text{seconds}$$

and

$$\text{watts} = \text{amperes} \times \text{volts}$$

(See *newton, joule, watt, ampere, coulomb, volt*.)

electrical circuit calculations are numerical exercises which involve the application of Ohm's law, or other circuit principles, to the calculation of an unknown circuit component or quantity. The relevant ideas are:

- Ohm's law: $E = IR$ applies to a full circuit.
  $V = IR$ applies to any component.
- electrical power: $P = IV = I^2R = V^2/R$
- series resistors formula: $R = R_1 + R_2 + R_2$
- parallel resistors formula:

$$\frac{1}{R} = \frac{1}{R_1} + \frac{1}{R_2} + \frac{1}{R_3}$$

- Kirchhoff's first law: $I = I_1 + I_2 + I_3$
- Kirchhoff's second law: $EI = I_1V_1 + I_2V_2 + I_3V_3$ for any complete loop in the circuit.

Skill with these laws comes with practice. Notice that Kirchhoff's first law and the parallel resistor formula apply whenever a circuit divides.

electrical power is either the rate of supply of electrical energy (= e.m.f. × current) or the rate of dissipation of electrical energy (= potential difference × current). The SI unit for electrical power is the watt for which we now have two possible word equations:

$$\text{watts} = \frac{\text{joules}}{\text{seconds}} \qquad W = J\,s^{-1}$$

$$\text{watts} = \text{volts} \times \text{amps} \qquad W = VA$$

electromagnetic induction is the setting up of an electromotive force (e.m.f.) in a conductor by a change of *magnetic flux linkage*. Two typical situations, illustrated in the diagram below, are a magnet falling through a coil and a conductor moving at right angles to a magnetic field. They represent two general cases: a fixed conductor in a changing magnetic field and a moving conductor in a fixed magnetic field. In neither case need the conductor be part of a complete circuit. If the circuit is incomplete there will be no current, the e.m.f. is unaffected (this is the same situation as a battery in a circuit with the switch left open).

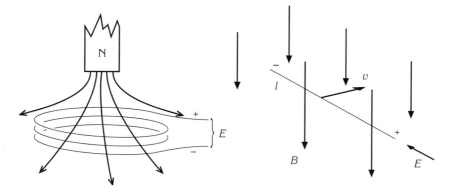

With a coil, let ϕ represent the instantaneous flux threading the coil and let N represent the number of turns in the coil. The number of *flux linkages* for the coil is $N\phi$. If $N\phi$ is changing, then the induced e.m.f. E equals the rate of change of flux linkages:

$$E = -\frac{d}{dt}(N\phi) \qquad = -NA\frac{dB}{dt}$$

where N is constant and ϕ equals BA and B is the average value across the coil of the magnetic field strength. The negative sign indicates that if ϕ is increasing then any current driven by E would create flux opposite in direction to ϕ, thereby decreasing the rate of change of flux. If ϕ were decreasing, then any current driven by E would create flux in the same direction as ϕ.

Faraday's law of electromagnetic induction states that induced e.m.f. is proportional to rate of change of flux. We seem to have missed out a constant of proportionality in the equation for E. There is a constant of proportionality but, because of the way electrical units are defined, it equals 1 and has no unit.

Another way of looking at the situation is to observe that for each complete turn in the coil, there is an induced e.m.f. E', where:

$$E' = -\frac{d}{dt}(\phi) \qquad = -A\frac{dB}{dt}$$

and since there are N turns in series, the total e.m.f is:

$$E = N \times E' \qquad = -N\frac{d}{dt}(\phi) \qquad = -NA\frac{dB}{dt}$$

In the case of a single wire of length l cutting across a magnetic field of uniform magnetic field strength B, the rate of change of flux is decided by the velocity v of the wire:

$$E = -\frac{d}{dt}(N\phi) \qquad = -B\frac{dA}{dt} \qquad = -Bl\frac{dx}{dt} \qquad = -Blv$$

The significance of the minus sign here is that any current created by the induced e.m.f. would have direction such that the force set up on it by the magnetic field would oppose v. The agent responsible for moving the wire would do work on the wire by keeping it moving.

Another way of looking at the minus sign is to recognise that the magnetic flux ϕ could well originate with a current l in the coil itself. Since ϕ would then be proportional to l, the equation for the induced e.m.f. could be written in the form:

$$E = -\text{constant} \times \frac{dl}{dt} = -L\frac{dl}{dt}$$

If the current l 'tries' to grow, the induced e.m.f. E and the current l will be in opposite directions and the growth of current will be inhibited. If the current l 'tries' to get smaller, the induced e.m.f. will be in the same direction as the current l and a fall in the value of l will be inhibited. The constant L is called the self-inductance of the coil.

The direction of the induced e.m.f. can be worked out also from Fleming's right hand rule.

There is no way of discovering that electromagnetic induction happens, except by experiment. But once it is known that it happens, the law of conservation of energy must apply. The laws of electromagnetic induction can then be derived analytically.

Where the induced e.m.f. drives an induced current, electrical energy is transformed into heat energy. In the case where the e.m.f. comes from a change of magnetic flux linkages in a coil, energy stored in a magnetic field is transformed first into electrical energy and then into heat energy. Where the e.m.f. derives from the motion of a conducting wire across a magnetic field, energy transferred by mechanical work is transformed first into electrical energy and then into heat energy.

(See electromagnetic induction (laws of).)

electromagnetic induction (laws of): there are two laws, a law named after Faraday which gives the magnitude of an induced e.m.f. and a law named after Lenz which gives its direction.

You need to remember that flux linkages equal $N\phi$ where N is the number of turns and ϕ equals the magnetic flux threading a circuit. For uniform magnetic fields, ϕ equals BA where B is the magnetic field strength and A is the cross-sectional area of the coil imagined to be at right angles to B.

Faraday's law:

The magnitude of the e.m.f. induced in a conductor equals the rate of change of flux linkages or the rate at which the conductor cuts a magnetic flux.

Lenz's law:

The direction of the e.m.f. induced in a conductor is such that it opposes the change producing it.

(The word 'equals' instead of 'proportional to' in Faraday's law comes about because of the way in which flux and flux linkage are defined, but this need not worry you. 'Conductor' is used instead of 'circuit' because an e.m.f. will be induced in a conductor if the circumstances are right whether or not it is part of a circuit. For the same reason there is no reference in the laws to induced current.)

An easier way to work out the direction of the induced e.m.f. is given by Fleming's right hand rule. (See *electromagnetic induction, Fleming's right hand rule, flux linkage*.)

electromagnetic spectrum: the range of electromagnetic waves listed in order of increasing wavelength and named in segments according to their mode of origin. The main segments of the electromagnetic spectrum are as follows: gamma radiation, X-rays, UV (ultra-violet), visible, IR (infra-red), microwaves, radio waves. Each of these segments is described in more detail under its own heading. The main points are listed in the table on page 78.

Notice that this table is not linear. The top of the X-ray energy band, 1.2 MeV appears to be halfway between the bottom of the gamma radiation band at 1.2 keV and the top of the gamma radiation band at 1.2 GeV. In fact the scale is logarithmic and 1.2 MeV is a thousandth part of 1.2 GeV. This logarithmic scale can be very misleading. The energy carried by radio-waves is minuscule compared with the energy carried by gamma radiation. Long-wave radio powered by an ordinary 9 V battery can be heard around the world.

The wavelength bands overlap because the radiation is named according to mode of origin and not wavelength. (See *gamma radiation, X-radiation, UV, visible light, IR, microwave and radar radiation, radio waves, photon*.)

electromagnetic waves are a flow of energy in the form of linked oscillating electric and magnetic fields. The energy is carried in discrete units known as *photons*. Each photon has energy hf where h is the Planck constant and f is the frequency of the electromagnetic oscillation. Electromagnetic waves are classified into different groups which are named according to the mode of origin. (See *electromagnetic spectrum*.)

electromotive force: see *e.m.f.*

electron: an elementary particle of rest mass $9.109\,389 \times 10^{-31}$ kg which carries a negative electric charge of $1.602\,177 \times 10^{-19}$ C. (See *leptons*.)

electron beams usually consist of *electrons* released from a hot cathode by *thermionic emission*, all moving through vacuum with the same acceleration from cathode to anode. The speed with which the electrons arrive at the anode is calculated by equating the mechanical kinetic energy gained to the electrical potential energy lost:

$$eV = \tfrac{1}{2}m_e v^2$$

where e is the electron charge, V is the potential difference responsible for the acceleration of the electron, m_e is the electron mass and v is the final speed of the electron. V is numerically equal to the energy of the electron in electronvolts.

1 electronvolt (1 eV) $= 1 \times (1.602 \times 10^{-19}$ C$) \times (1$ volt$) = 1.602 \times 10^{-19}$ J

The electromagnetic spectrum

Wavelength	Frequency	Photon energy	Name	Origin	Detection	Properties	Use
1 fm – 1 nm	0.3 YHz – 0.3 EHz	1.2 GeV – 1.2 keV	Gamma radiation	Nuclear decay	Scintillation counter GM tube (poor) Photographic film	High penetration Very high photon energy Damaging	Medical diagnosis Tracing underground pipes
1 pm – 10 nm	0.3 ZHz – 30 PHz	1.2 MeV – 120 eV	X–rays	Inner shell excitation X–ray tube	GM tube Photographic film	High penetration of low Z materials	Diagnosis (low energy) Therapy (high energy) Crystallography
1 nm – 400 nm	0.3 EHz – 0.75 PHz	1.2 keV – 3.1 eV	UV Ultraviolet	Inner shell excitation Mercury vapour lamps Carbon arcs	Photocells Photographic film Solid-state detectors Fluorescence	Damages living cells Absorbed in stratosphere Chemically destructive	Sterilising food etc. Investigating atomic structure Invisible labelling
400 nm – 700 nm	0.75 PHz – 0.43 PHz	3.1 eV – 1.8 eV	Visible	Outer shell excitation Hot bodies	Eyes Photocells Photographic film	Selective absorption by coloured surface Chemically active	Vision Photosynthesis Communication
700 nm – 1 mm	0.43 PHz – 0.30 THz	1.8 eV – 1.2 meV	IR Infra-red	Molecular excitation Hot bodies	Thermopiles Photography, photo cells for near IR	Penetrates fog and mist, Emitted by dark surfaces	Satellite reconnaissance Molecular structure IR control for TV etc.
1 mm – 10 cm	0.30 THz – 3.0 GHz	1.2 meV – 12 μeV	Micro-waves	Klystron or magnetron valves	Tuned cavities Point contact diodes	Focuses into beam Absorbed by water Reflected by metals	Microwave cooking Communication Satellite reconnaissance
> 10 cm	< 3.0 GHz	< 12 μeV	Radio	Electrons accelerated in metals (aerials)	Receiving aerials attached to tuned circuits	Reflected by ionosphere Strongly diffracted	Communication and navigation systems Radioastronomy

Powers of ten: Y→24 Z→21 E→18 P→15 T→12 G→9 M→6 k→3 m→-3 μ→-6 n→-9 p→-12 f→-15 a→-18 z→-21 y→-24

The *electronvolt* may seem an odd unit of energy but it is very useful in atomic and nuclear physics.

The energy equation is often used in the alternative form:

$$v = \sqrt{\left(\frac{2eV}{m}\right)}$$

Electron beams with all the electrons moving at this same constant speed can be produced by arranging for the electrons to pass through a hole in the anode. This is the principle of the electron gun. Often, there are two cylindrical anodes held at slightly different voltages which both accelerate the electrons and focus them into a narrow beam. Electron beams can be deflected by electric or magnetic fields. An electron which is moving through an electric field and parallel to the lines of force will be accelerated; this is how electron beams get their energy. A magnetic field which is parallel to the direction of electron motion has no effect.

The diagram below shows three other common situations. Electrons, all with speed v, moving through a uniform electric field at right angles to the direction of motion retain their forwards speed v but accelerate against the direction of the field (against – because of their negative charge). In a similar way to a football kicked sideways over a cliff, the electrons follow a parabolic path. The parabola is tightened by either reducing the speed v or increasing the electric field strength.

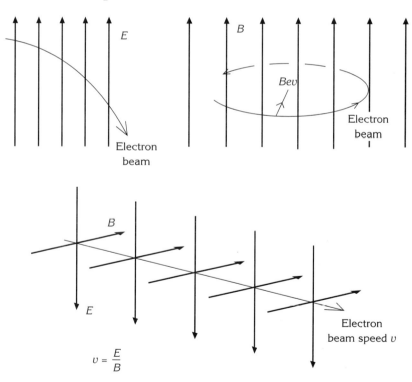

Electrons moving with constant speed v at right angles to a magnetic field of strength B are subjected to a force of magnitude Bev directed at right angles both to the the path of the electron and to the magnetic field. The actual direction satisfies *Fleming's left hand rule* – but remember that the current direction is opposite to the direction in which the electrons are moving. As the electron beam swings round, the direction of the deflecting force swings round too. The deflecting force is a centripetal force and the electrons follow a circular path. We may write:

$$Bev = \frac{mv^2}{r} \qquad \text{or} \qquad r = \frac{mv}{Be}$$

The third important case is shown in the third part of the diagram; an electron beam with all the electrons moving at constant speed *v* at right angles to a magnetic field of strength *B* and an electric field of strength *E* with both *B* and *E* themselves at right angles. The magnitudes and direction can be arranged for the two deflections to just cancel and for the electrons to continue along their straight line paths. For this to happen the speed of the electrons must be just right:

$$Bev = eE \qquad \text{or} \qquad v = \frac{E}{B}$$

This may seem like a non-event. In fact, if the electrons enter the crossed fields at all sorts of speeds and directions, only electrons with the right speed and the right direction will get through. The arrangement is called a velocity selector. The principle of the velocity selector can be extended for use with any charged particle, especially protons and ions.

electron diffraction is illustrated by passing a narrow beam of electrons through thin gold foil. A detecting screen beyond the gold foil shows that the electrons, when they arrive, form a pattern of rings centred on the spot towards which the electrons were directed. The ring pattern is much the same as a diffraction pattern set up by a laser passing through a small circular hole. The relevant wavelength for the electron is the *de Broglie wavelength λ* for the electron and is given by the formula:

$$\lambda = \frac{h}{m_e v}$$

where *h* is the Planck constant, m_e is electron mass and *v* is the speed of the electron.

If the electron wavelength is calculated and the radius of the diffraction rings measured, then a typical dimension can be calculated for the diffracting object. For low energy electrons (about 100 V), information is collected about atomic size and crystal structure. Energies of up to 100 MeV are used to examine nuclear size and electrons with energies of around 100 GeV are used to penetrate the proton and gain a modest amount of information about proton structure and quark diameter (less than 10^{-18} m). Protons and other subatomic particles can be used in a similar way. See *diffraction at a small circular hole*.

electron volt: the energy transferred to a free electron when it crosses a potential difference of 1 volt. Using the relation *W = QV* we find that:

$$1 \text{ eV} = (1.602 \times 10^{-19} \text{ C}) \times (1 \text{ V}) = 1.602 \times 10^{-19} \text{ J}$$

It is a useful energy unit when discussing events at the atomic level.

electronic charge is the charge carried by an electron. It is written:

$$e = -1.602\ 177 \times 10^{-19} \text{ coulombs}$$

electrostatics is the study of electric fields, charged matter and their interactions.

e.m.f. (electromotive force) is the energy transformed into electrical energy per coulomb of charge flowing in a circuit. The symbol used is E and the SI unit is the volt (V).

There is a real and vital difference between e.m.f. and *potential difference*. In loose terms, e.m.f. is the energy supply which makes current flow and potential difference is the energy dissipated (often as heat) when current flows. A definition for electrical potential difference which would match the above definition of e.m.f. is:

'the potential difference between two points is the electrical energy transformed per coulomb of charge flowing between the two points.'

An understanding of the difference between e.m.f. and potential difference is fundamental to an understanding of *Kirchhoff's second law*.

With the increasing use of 'e.m.f.' as a noun in its own right, there is less insistence on using the full but cumbersome 'electromotive force'.

emission spectrum: the electromagnetic radiation emitted by a hot body spread out to show how intensity varies with wavelength. It may be continuous as in the *total radiation curve* or discrete as in the line spectrum of an element such as hydrogen or a band spectrum from a compound such as water. Line spectra have wavelengths ranging from the X-ray region to the IR, according to which electrons in which particular atoms are excited. See *absorption spectrum*.

end states of stars: there are three end states, namely: white dwarfs, neutron stars and black holes. Red giants and supernovae are intermediate stages.

1 Red giants and white dwarfs. If the mass of a main sequence star is less than eight solar masses (8 M_{Sun}) then, once the core hydrogen is largely used up, the hydrogen in the outer shell starts to fuse and the radiation pressure set up causes the radius of the outer shell of the star to increase a hundred-fold and the surface area ten thousand-fold. The rate of energy release by fusion will increase a thousand-fold but the surface area increases ten times as much, so the surface brightness diminishes. The surface is cooler and the radius much larger, hence the term red giant.

After about 10^9 years in the red giant state, the helium in the core suddenly starts to fuse. The whole process is completed in a matter of seconds. The sudden outburst of energy throws the shell away from the core. The core, robbed of any energy releasing process, shrinks into a high density state in which further collapse is prevented by pressure generated in the electron gas and known as '*electron degeneracy pressure*'. The high temperature causes the star to look white and the small radius (typically, the mass of the Sun in the volume of the Earth) explains the description 'dwarf', hence, white dwarf. The white dwarf cools down over about 10^9 years and becomes a black dwarf. All this can only happen if the mass of the star remnant is less than 1.4 M_{Sun} or main sequence mass less than 8 M_{Sun}.

2 Supernovae and neutron stars. Once the core hydrogen of a larger star is mostly gone, the core contracts and cools. If total mass of the star is greater than about 8 M_{Sun} then the mass of the core is likely to exceed 1.4 M_{Sun} (known as the Chandrasekhar limit). In these circumstances, nuclei suddenly break up into a neutron–proton–electron mix. The protons released combine with electrons to give a neutron core which collapses in the absence of electromagnetic repulsion. Contraction is catastrophic, an enormous amount of gravitational energy is released through the shell and this energy flow (a shock wave) prompts the immediate fusion of much of the shell material. For a short while (about a year) there is a spectacular release of radiant energy known as a supernova. The core is left behind as a neutron star whose mass is typically a few solar masses, with a radius equal to about ten kilometres. The density is enormous. The small radius implies a much reduced moment of inertia and a high spin speed (in radians per ms). This in turn explains the high repetition rate of the bursts of radio waves from pulsars – which are thought to be neutron stars.

3 **Black holes.** These form if the remnant from a supernova has mass greater than about 2.5 M_{Sun}. This happens if the original total mass of the star exceeded about 20 M_{Sun}. The neutron core contracts into a radius about three or four times smaller still (i.e. a radius of a few kilometres instead of ten kilometres). The gravitational field near the surface is so strong that not even light can escape, . This is why they are called black holes. Efforts to detect black holes concentrate on either the possible influence on a nearby object if the black hole is part of a binary system or on the deflection of a beam of light from an object far behind the black hole.

Another definition of a black hole is a star whose density is so great that the escape velocity exceeds the speed of light. The escape velocity v for a star (or planet) of mass M is $(2GM/R)^{1/2}$ so the radius of a black hole of mass equal to the sun's mass M_s is $M_s.(2G/c^2)$. This is called the Schwarzschild radius. For the Sun, it comes out at 2.95 km. See *Hertzsprung–Russell diagram*, *star*.

energetics of stretching: an examination of what happens to the energy transferred to a material when it is stretched. There are three possibilities:

- a material may be stretched elastically in which case the strain energy is recoverable
- a material (such as rubber) may exhibit a hysteresis effect with only part of the energy recoverable
- a material may pass its yield point in which case plastic flow will ensue and a proportion of the strain lost, dissipated in breaking down the structure of the material.

(See *stress–strain curves*.)

energy is the conserved quantity referred to in the *first law of thermodynamics*. It is a scalar quantity. The SI unit of energy is the joule, (J). The usual symbol is W but, there are other symbols such as E_k for kinetic energy.

The energy concept is now a fundamental part of our thinking. Energy is an intangible quantity. It cannot be seen or observed directly. An amount of energy is always calculated from other quantities which can be observed and measured. Almost all physical processes can be analysed as energy transfer or energy conversion processes. (See also *work*, *internal energy*.)

energy level: a number of joules of energy which can be acquired and stored by an extra-nuclear electron, initially in the ground state, without the atom's hold on the electron becoming unstable.

An atom which absorbs energy and stores it by transferring an electron to an energy level above the ground state is said to be in an 'excited' state. An atom which absorbs so much energy that an electron is ejected from the atom is said to be in an ionised state. Ionised atoms carry net positive charge, excited atoms do not.

The word 'state' has to be used carefully. It means the set of physical quantities such as energy, angular momentum and spin which together define the condition of an electron within an atom. The minimum set of values is called the ground state. Other states, the excited states, are defined with reference to the ground state.

If an atom transfers spontaneously from an excited state with energy E_1 to a lower state of energy E_2, the energy $E_1 - E_2$ is released in the form of a photon of frequency f, where $E_1 - E_2 = hf$. By measuring the energy changes corresponding to all the spectral lines emitted by an atom, it is possible to map the energy levels.

energy–separation curve: a graph which shows how the energy stored in the bond between two atoms varies with their separation. The diagram shows a typical example. The part of the curve where the bond energy is negative corresponds to the stable states; energy is needed to separate the atoms from one another.

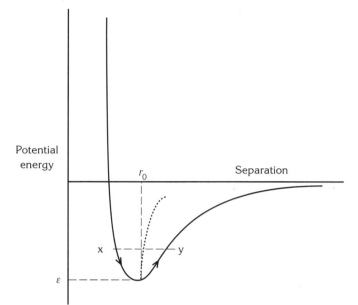

At absolute zero r_0 is the separation of the centres of the two atoms. It is approximately equal to the atomic diameter and can be anything from about 0.1 nm to 0.5 nm. Above absolute zero, the atomic pair has more than this minimum amount of energy and the extra energy allows the atomic pair to oscillate between a minimum separation at X and a maximum separation at Y. The average separation is found on the dotted line which creeps upwards and outwards. The steepness of this dotted line is proportional to the linear expansivity of the material. A deep potential well implies low linear expansivity, high specific latent heats, high melting and boiling points, high *Young's modulus* and high *strength*.

energy sources are the fuels needed by national economies. The *primary fuels* are oil (40%), coal (30%), natural gas (20%) and the remaining 10% almost equally split between nuclear sources and hydroelectricity. These figures are proportions that vary from one country to another and change slowly from year to year. A more worrying aspect is the total. The rate of fuel use has increased roughly tenfold over the last hundred years (about three times as fast as population growth). A similar rise in the next century would be a hundred-fold increase in 200 years. If this were to be provided by a redirection of solar energy, it might be tolerable. If it were to come from a similar increase in the rate of burning of fossil fuels it must surely prove disastrous.

Primary fuels are in their natural state; they are either finite or renewable. Other than nuclear fuels, finite fuels are mostly fossil fuels. Renewable fuels include solar energy, hydroelectricity, wind power and biofuels. The term *biofuels* is of recent origin and should not be extended to include *fossil fuels*. Peat is a fossil fuel, wood a biofuel. *Secondary fuels* such as electricity or petrol are produced from primary fuels. (See also *finite energy sources, fissile materials, renewable energy sources, thermal power station, thermal fission reactor.*)

energy stored in a capacitor of capacitance C and with a potential difference V across its plates is ½CV². To derive this result, we look at the graph of potential difference V (ordinate) against charge Q (abscissa). The diagram below of the capacitor shows the negative plate earthed to ensure that it remains at a zero reference potential.

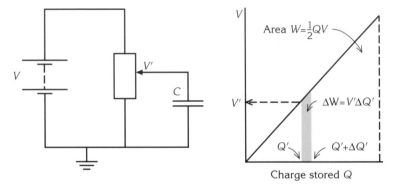

Charge stored Q

At some point in the charging process, the potential difference across the capacitor will be V' and the charge stored will be Q'. Imagine a small extra charge +ΔQ' added to the positive plate. An equal but opposite small negative charge will be drawn from earth to the earthed plate to keep it at earth potential. The work expended in bringing the charge +ΔQ' to the positive plate is + V'ΔQ' and this equals the area of the shaded strip of width ΔQ' and (approximate) height V'. It follows that the work W expended in bringing the whole charge +Q to the positive plate is the sum of the areas of countless thin strips, i.e. the area between the potential-charge line and the charge axis. Therefore the energy W stored in a charged capacitor is ½QV.

$$Q = CV \quad \text{and} \quad W = \frac{1}{2}QV = \frac{1}{2}CV^2 = \frac{Q^2}{2C}$$

Charging current

Heat dissipated = $\frac{1}{2}QV$

Energy stored = $\frac{1}{2}QV$

Energy transferred
= QV

The symbol W in the previous paragraph represents the energy stored in the charged capacitor. The energy transferred from the battery during the charging process is different. The simplest way to charge a capacitor is to connect it across a battery, through a resistor to control the current. The situation is illustrated in the diagram. The charge Q which ends up on the capacitor leaves the battery at potential V relative to earth and carries electrical energy QV from the battery. Only half of this energy is stored in the capacitor. The other half is expended in the resistor by the charging current. From the battery's 'point of view', the energy stored in the capacitor is literally only half of the story.

Worked example

Calculate the energy stored in the 4 µF capacitor in the circuit below once the charging process is complete.

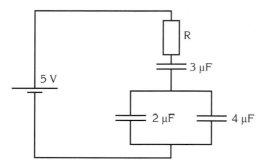

To calculate the energy stored in a known capacitance, the voltage across it must be known. To find this we find the charge on the 3 µF capacitor, but first we need to know the equivalent capacitance of the whole arrangement. The 2 µF in parallel with the 4 µF are equivalent to 6 µF (see *capacitors in parallel*). The equivalent capacitance for 6 µF in series with 3 µF is calculated as follows:

$$\frac{1}{C} = \frac{1}{C_1} + \frac{1}{C_2} = \frac{1}{3} + \frac{1}{6} = \frac{1}{2}$$

$$C = 2 \text{ µF}$$

The charge carried by this equivalent capacitance is found from the relationship $Q = CV$. With V equal to 5 V, the charge is 10 µC. This equals the charge on the 3 µF capacitor. The voltage across the 3 µF capacitor is found from $V = Q/C$. It equals 3.33 V. The voltage across the 4 µF capacitor must be 1.67 V and the energy stored is 5.6 µJ.

energy transfer is either the transfer of energy from one body to another or the displacement of energy within a body. The transfer of energy in an ordered form is called *work*. The transfer of energy in the disordered form is called *heating*. (See *heating and working*.)

engine cycle: the full cycle of actions undergone by a repetitive engine such as a steam engines or a car engine. A cyclical engine is in exactly the same condition at the end of a cycle as it was in when the cycle began. Input and output can be precisely balanced.

The equation for the efficiency of a *heat engine*, $\eta = [T_1 - T_2]/T_1$, imposes an upper limit to the efficiency of any engines which transforms heat into mechanical energy, whether or not it is cyclical. It applies, for instance, to a thermocouple delivering current to a micro-motor or to the human body drawing mechanical energy from its food supply. The derivation of the equation hinges on the first and second laws of *thermodynamics*.

Consider two engine cycles, the four stroke petrol cycle and the Diesel engine cycle. The main difference between them is the ignition phase which is triggered by a spark in the four stroke petrol engine but is spontaneous in the Diesel engine. This is because the Diesel engine works at a significantly higher temperature and pressure. The two engine cycles are illustrated in the diagram. Such diagrams are, for historical reasons, called indicator diagrams. The area enclosed by the loop in an indicator diagram equals the work done by the engine per cycle.

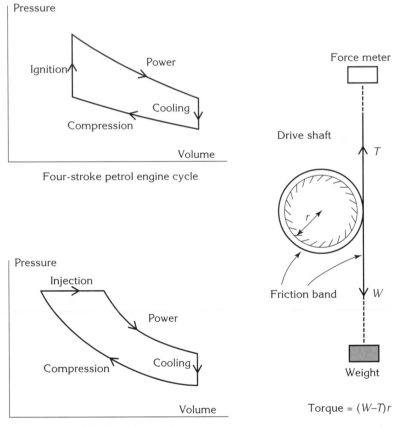

Four-stroke petrol engine cycle

Diesel engine cycle

Torque = $(W-T)r$

The theoretical upper limit to the efficiency of the four stroke engine is about 60% but in practice it is about 30%. The Diesel engine runs at a slightly higher temperature, there is a higher theoretical efficiency and the efficiency in practice is nearer 40%. These efficiencies can be checked by running the engine against a frictional torque applied to the drive shaft by a large mass of known weight supported by a friction band. The friction band runs once round the shaft and the difference in tensions $(T - W)$ multiplied by the radius of the shaft is the output torque. The output energy, equal to the output torque multiplied by the angle turned in radians, can be compared with the energy provided by the fuel.

enthalpy H is the sum of the internal energy U of a body and the product of the pressure p and the volume V:

$$H = U + pV$$

A change of enthalpy at constant pressure has the form:

$$\Delta H = \Delta U + p\Delta V$$

If this equation is compared with the *first law of thermodynamics*, it is seen that a change of enthalpy is the heat exchanged with the surroundings in conditions of constant pressure.

equation: a statement that two quantities are equal to one another. The equality must relate both to magnitude and to the kind of quantity involved. An equation must have an equals sign, not a proportionality sign nor the arrow used in chemical or nuclear reaction statements.

Equations arise in different ways. Sometimes (as in the equation $a = (v - u)/t$) the right-hand side defines the quantity on the left-hand side.

Sometimes (as in the equation $s/t = (v + u)/2$) the two sides are different expressions of the same quantity – average velocity in this example.

Sometimes the equation makes a physical statement. The equation $I = nAvq$ states that current is rate of flow of charge.

Sometimes, as in $pV = nRT$, the equation is a relationship between different quantities and awaits explanation.

It is worth developing the habit of saying to yourself when you meet a new equation 'What is this equation actually saying?'.

equations of rotational motion follow the same pattern as the equations for straight line motion but with angle, angular velocity and angular acceleration replacing displacement, velocity and acceleration.

There are five variables:

θ – angular displacement in radians (θ must be zero at time $t = 0$)

t – time in seconds

ω_1 – initial angular velocity (i.e. at time $t = 0$) in rad s^{-1}

ω_2 – final angular velocity (i.e. at time t) in rad s^{-1}

α – angular acceleration (which will be constant or zero) in rad s^{-2}.

The angular displacement, θ, has positive and negative directions. The positive direction for angular displacement, which must be chosen and then kept to, sets the positive direction of rotation for angular velocity and for angular acceleration. The initial and final angular velocities and the angular acceleration, like angular displacement, may have positive or negative values. This underlines their vector nature. They are *pseudo-vectors* in the sense that their directions are a matter of convention. Only t is a scalar. Negative values of t would imply times earlier than $t = 0$.

Notice that each of the equations has one of the five variables missing. The equations of rotational motion are:

$$\theta = \frac{\omega_1 + \omega_2}{2} \times t \qquad \alpha \text{ missing}$$

$$\omega_1 = \omega_2 + \alpha t \qquad \theta \text{ missing}$$

$$\theta = \omega_1 t + \frac{1}{2}\alpha t^2 \qquad \omega_2 \text{ missing}$$

$$\omega_2^2 = \omega_1^2 + 2\alpha\theta \qquad t \text{ missing}$$

The first equation follows from the definition of average angular velocity. The second equation is the definition of uniform angular acceleration. The third is derived by eliminating ω_2 from the first two equations. The fourth equation is derived by eliminating t.

Any problem in rotational motion can be tackled by following a similar procedure to that for straight line motion by using:

moment of inertia I	instead of	mass m
resultant torque T	instead of	resultant force F
$T = I\alpha$	instead of	$F = ma$
$\frac{1}{2}I\omega^2$	instead of	$\frac{1}{2}mv^2$
$T \times \theta = \frac{1}{2}I\omega^2$	instead of	$F \times s = \frac{1}{2}mv^2$
power $= T \times \omega$	instead of	power $= F \times v$
$I\omega$ instead of	mv	
conservation of $I\omega$	instead of	conservation of mv
and angular impulse $= T \times t$	instead of	impulse $= F \times t$.

Never lose sight of the relevant direction of rotation.

See *moment of inertia, angular displacement, angular speed, angular velocity, angular acceleration, angular momentum, torque, couple, moment of a force*.

equations of straight line motion are the five equations which describe uniformly accelerated motion along a straight line. There are five variables involved:

s – displacement in metres (s must be zero at time $t = 0$)

t – time in seconds

u – initial velocity (i.e. velocity at time $t = 0$) in metres per second (m s^{-1})

v – final velocity (i.e. velocity at time t) in metres per second (m s^{-1})

a – acceleration (which must be constant or zero) in metres per second squared (m s^{-2})

Since s is displacement, not distance or distance travelled, the straight line has positive and negative directions. The positive direction for displacement sets the positive direction for velocity and for acceleration. The initial and final velocities and the acceleration, like displacement, may have positive or negative values. This underlines their vector nature. Only t is a scalar. Negative values of t imply times earlier than $t = 0$.

Notice that each of the five variables is missing from one of the equations. The equations of motion are:

$$s = \frac{u + v}{2}t \qquad \text{(a missing)}$$

$$v = u + at \qquad \text{(s missing)}$$

$$s = ut + \tfrac{1}{2}at^2 \qquad \text{(v missing)}$$

$$s = vt - \tfrac{1}{2}at^2 \qquad \text{(u missing)}$$

$$v^2 = u^2 + 2as \qquad \text{(t missing)}$$

The first equation follows from the definition of average *velocity*. The second equation is the definition of uniform *acceleration*. The last three equations can be derived algebraically from the first two; the third by eliminating *v*, the fourth by eliminating *u* and the fifth by eliminating *t*.

It is instructive to see how the first four equations can be read out from a velocity–time graph.

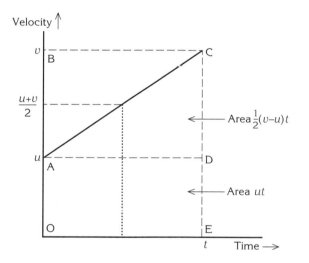

For the first equation, displacement s at time t equals the area OACDE and this equals half of (u + v) multiplied by OE, (i.e. multiplied by t) we have: s = [(v + u)/2] × t.

For the second equation, notice that acceleration a is the gradient of the graph, a equals CD/t, i.e. CD equals at. But CD = v − u, hence the second equation, v − u = at.

For the third equation, displacement s equals the area OACDE. But area OACDE = area OADE (=ut) plus the area ACD. The area ACD equals ½ AD times CD. But AD is t and CD is at. Hence the area ACD = ½ at². Whence s = ut + ½ at²

For the fourth equation, displacement equals area OBCE minus area ABC, i.e. vt − ½ at².

Worked examples

1. A student with a stopwatch drops a stick into a well and hears the plop 3.2 seconds later. Calculate the depth of the well and the final velocity of the stick. State two reasons why your answer for the depth of the well is an over-estimate.

Data:

t = 3.2 s, u = 0 m s^{-1}, a = 9.8 m s^{-2},   find s and then v.

s = $ut + \frac{1}{2}at^2$

= $0 + 0.5 \times (9.8$ m s$^{-2}) \times (3.2$ s$)^2$ = 50.18

= 50 m to 2 s.f.

v^2 = $u^2 + 2as$

= $0 + 2 \times (9.8$ m s$^{-2}) \times 50.18$ = 983.4

v = $\sqrt{983.4}$ = 31.36

= 31 m s^{-2} (to 2 s.f.)

Notice that the use of *quantity algebra* ensures that only SI values go into SI equations. Notice, too, that the calculations are worked through with as many significant figures as appear during the calculation but the extra significant figures are removed from the answers.

If there were no air resistance t would be smaller, if eyes were used in place of ears, t would be smaller because light travels so much faster than sound (it would be smaller by about 0.15 second). The exaggerated fall time gives an exaggerated depth.

2. An aeroplane in level flight speeds up from 323 km h^{-1} to 476 km h^{-1} in 5.81 minutes. How far does it travel in this time.

The first step must be to list the relevant data in SI units (no multiples or submultiples!)

Data:

t = 5.81 min = 348.6 seconds, u = 89.72 m s^{-1}, v = 132.2 m s^{-1}

$$s = \frac{u + v}{2} \times t$$

$$s = \frac{(132.2 \text{ m s}^{-1}) + (89.72 \text{ m s}^{-1})}{2} \times (348.6 \text{ s})$$

= 38 680.66 = 38.7 km   to 3 s.f.

The equations of motion are derived from the definitions of average velocity and uniform acceleration. Each equation contains four variables; you must always be given data for three of them. Part of the data is often given in a hidden form. One way of doing this is to give a mass and a force from which uniform acceleration must be derived. Another way is by employing potential energy lost equals kinetic energy gained to find, say, a final velocity value. Alternatively, a word like 'dropped' implies both that the initial velocity is zero and that the acceleration is that of free fall.

equilibrant is the name given to the single vector which is of equal magnitude to the resultant of two or more vectors, but which points in the opposite direction. It balances the resultant. For example, the tension T in the chain shown in the diagram on the next page is the equilibrant of the weight W of the bucket and the sideways force F. In fact, when three or more forces are in equilibrium, any one force is the equilibrant of the remaining forces. (See *addition of vectors*.)

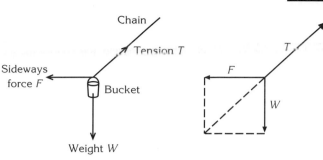

equilibrium: see *forces in equilibrium*.

error is the difference between the result of a measurement and the true value. An error cannot be found unless the true value is known. Where the true value is not known, the relevant concept to deal with is *uncertainty*. Some people use the term 'probable error' instead of 'uncertainty'. Errors can be stated in absolute terms, e.g. 4 mm, or as a percentage of the true value, e.g. – 5%. Notice that the symbol ± is used with uncertainties but not with errors. Either an error is positive or it is negative. (See also *random error, systematic error, precision, sensitivity*.)

escape speed is the vertical speed with which an object must leave the Earth's surface (or the surface of some other planet or star) to escape from the gravitational field. To do this, it must have more kinetic energy than the potential energy it would lose were it to fall to Earth from an infinite distance away.

The gravitational potential near the Earth's surface is $- GM/r$ where G is the universal gravitational constant (6.67×10^{-11} N m^2 kg^{-2}), M is the Earth's mass (5.98×10^{24} kg) and r is the Earth's radius (6.37×10^6 m). The potential energy of an object of mass m at the Earth's surface is $- GMm/r$, and for the object to escape from the Earth's gravitational field its total energy within the field must be positive, that is:

$$\tfrac{1}{2}mv^2 - GMm/r > 0$$

$$v > \sqrt{\left(\frac{2GM}{r}\right)}$$

$$v > \sqrt{\left(\frac{2 \times (6.67 \times 10^{-11} \text{ N m}^2 \text{ kg}^{-2}) \times (5.98 \times 10^{24} \text{ kg})}{(6.37 \times 10^6 \text{ m})}\right)}$$

$$v = 11.2 \times 10^3 \text{ m s}^{-1} \quad \text{(to 3 s.f.)}$$

(See *gravitational potential (radial fields)*.)

evaporation and cooling are processes which occur together. The average kinetic energy of the molecules in a liquid mass depends only on its temperature. But there is a spread of energy values among the molecules (see *distribution of speeds*). Molecules which are both near the surface and moving faster than average can escape from the liquid mass. This is evaporation. But the molecules involved in evaporation take away from the liquid more than their share of molecular kinetic energy. The average kinetic energy of the remaining molecules is therefore smaller than before and the temperature goes down.

exchange particles are formed momentarily when subatomic particles react under the influence of the *four fundamental interactions*. They exist momentarily during the interaction and their existence is essential to whatever process is going on. Their existence is made possible by the Heisenberg uncertainty principle which asserts that a particle of mass–energy ΔE can exist spontaneously for time Δt provided that $\Delta E. \Delta t \leq h / 2\pi$ where h is Planck's constant. The higher the mass-energy of an exchange particle, the shorter the lifetime.

Exchange particles come in four sets, one set for each fundamental interaction, and they are collectively called gauge bosons. Bosons are particles with integer spin; they are not subject to the Pauli exclusion principle. Gauge bosons are called virtual particles in recognition of their dependence on Heisenberg's uncertainty principle.

Photons are associated with the electromagnetic interaction acting alone or with the electroweak interaction. They are formed when charged particles either collide with one another or are deflected in a strong electric or magnetic field. Examples range from radio waves emitted from aerials, through visible photons released when excited atoms return to a lower energy state, to high energy X-ray photons associated with charged matter racing into a black hole from a nearby star.

Three exchange particles are associated with the weak interaction, namely the W^-, W^+ and Z^0 bosons. Their masses are roughly a hundred times the proton mass and their lifetimes are about 10^{-25} s. This does not seem very long but it is long enough for reactions that take place at the scale of a quark. Following this short lifetime, exchange particles decay into leptons and neutrinos. W particles, which come about when neutrons decay into protons or vice versa, themselves decay into the beta particles and neutrinos apparent in beta decay. (See *Feynman diagrams*.) The electroweak interaction comes about when electromagnetic forces are operating at the same time as the weak interaction.

Gluons are the exchange particles responsible for the operation of the strong interaction. There are eight kinds in all. (See *hadrons*.)

General relativity predicts the existence of gravity waves when two large masses interact gravitationally. Gravitons, which have yet to be observed, are introduced as part of an attempt to reconcile general relativity with the four fundamental interactions.

excitation is the process by which an atom receives energy and moves from the ground or non-excited state to a higher *energy level* or, as it is also called, an excited state.

experiments often have to be described within an A level question and the time allowed is usually about ten minutes. In general, your answer should be written out in three parts.

First, there should be a list of the necessary apparatus, almost always in the form of a labelled diagram.

Secondly, you should list the measurements that are made, which instruments are used and what is varied between successive measurements. Remember to stress that all measurements are repeated so that average values can be calculated for each value. If this cannot be done, as in a cooling experiment, for example, then stress that you are taking as many separate measurements as possible.

Lastly, explain how the conclusion is reached. This usually involves stating the variables on the two axes of a graph, what kind of graph is expected and how it is used to yield the result, e.g. by being a straight line passing through the origin or by having a gradient equal to whatever.

Do not be afraid to write out any of this in the form of a list. Your examiner will probably prefer it. Whether or not you have enough time to discuss precautions to eliminate errors depends on the wording of the question.

When time is limited, there is no point in providing extra detail or discussion. The mark scheme will not have an allowance to reward it. An experiment that would not work will get no credit, nor will an account of the wrong experiment.

exponential decay happens when the rate of decay of a quantity is a constant fraction of the present value of the quantity. Imagine you have £100 in a failing bank which reduces your balance by 15% at the end of each year. You will have £85 after one year, £72 after two years, then £61, £52, £44 and so on.

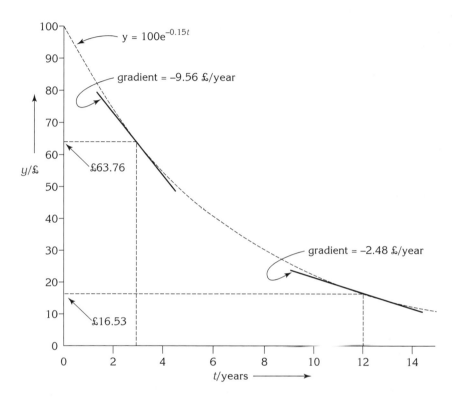

If the negative interest rate, still 15% per annum, is subtracted continuously instead of annually, then the balance would be more difficult to work out. If y represents the balance, then it would decay with time according to the following formula:

$$y = 100e^{-0.15t}$$

where the number e is called the exponential constant; it equals 2.71828 to six significant figures. A graph of balance y against time t in years is shown in the diagram above. It represents exponential decay with a halving time of 4.62 years and covers a little more than three half lives.

This graph is called an exponential decay curve. Its most important property is illustrated by evaluating the gradients of the tangents to the curve at two points chosen more or less at random. When t is 3 years, y is 63.76 and the gradient is –9.56. When t is 12 years, y is 16.53 and the gradient is –2.48. In both cases we have:

gradient = –0.15y

A quantity decays exponentially if the instantaneous decay rate equals a constant times the value of the quantity at that instant.

The following examples of exponential decay all follow a graph of the same shape as the curve in the diagram below but the axes would be labelled differently.

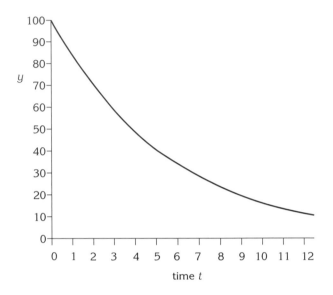

1 Damped harmonic motion. If the y-axis represents the peak value at an instant t of the amplitude x_0 associated with an object undergoing simple harmonic motion and if this peak amplitude diminishes at a rate proportional to the existing value of the peak amplitude, then:

$$\frac{dx_0}{dt} = -kx_0$$

where k is a constant. This leads us to the exponential decay of peak amplitude:

$$(x_0)_t = x_0 e^{-kt}$$

and the full equation for the displacement of an object undergoing damped harmonic motion becomes:

$$x = x_0 e^{-kt} \sin \omega t$$

2 Capacitor discharging through a resistor. If the y-axis represents the charge Q stored in a capacitor C and discharging through a resistance R, then the rate of discharge dQ/dt is proportional to the current I in the resistor. But the current is proportional to the voltage V across the capacitor, and this voltage is proportional to the charge on the capacitor. This leads us to:

$$\frac{dQ}{dt} = I = \frac{V}{R} = -\frac{Q}{RC} = -\frac{1}{RC}Q$$

The charge stored in the capacitor diminishes exponentially with time and the rate at which it diminishes is inversely proportional to the product RC.

3 Radioactive decay. If the y-axis represents the number N of radioactive nuclei in a sample still not decayed after time t, and if the number of nuclei decaying per second is proportional to N, then:

$$\frac{dN}{dt} = -\text{constant} \times N = -\lambda N$$

$$N = N_0 e^{-\lambda t}$$

The first equation is the fundamental law of radioactive decay, the second is the exponential law of radioactive decay.

All three examples follow the same mathematical pattern, they all lead to a graph of the same shape and all end up with an exponential decay law.

exponential law of radioactive decay states that if N represents the number of undecayed atoms of a particular nuclide present in a sample at time t, and if N_0 represents the number of undecayed atoms that were present at time $t = 0$, then:

$$N = N_0 e^{-\lambda t}$$

The *decay constant* λ is defined in the *fundamental law of radioactive decay*:

'The decay constant λ for a radioactive isotope is the ratio of the rate of decay for a large and pure sample of the isotope to the number of undecayed nuclei in the sample.'

Decay constant is shown below to be related to half-life as follows:

$$t_{1/2} = \frac{\ln 2}{\lambda} = \frac{0.693}{\lambda}$$

The exponential law derives from the fundamental law of radioactive decay but with the random character of the decay smoothed away. Imagine a sample of N nuclei, none of which has yet decayed. Provided that N is a very large number, the number ΔN which will decay in the next Δt seconds is proportional both to the number N of nuclei available for decay and the length of the time interval Δt:

$$\Delta N = -\lambda N \Delta t$$

where λ is a proportionality constant called the decay constant. It has units seconds $^{-1}$. This equation is called the fundamental law of radioactive decay. The negative sign is present because ΔN represents a reduction in the value of N. Both ΔN and Δt are finite quantities.

If N were a continuous variable instead of always being a whole number, the fundamental law of radioactive decay could be written in the modified form:

$$\frac{dN}{dt} = -\lambda N$$

This equation can be integrated between time zero and time t to give the exponential law of radioactive decay:

$$N = N_0 e^{-\lambda t}$$

The exponential decay curve is shown in the diagram on page 96. The half-life, the time interval over which half the specimen decays is the same for all points along the curve.

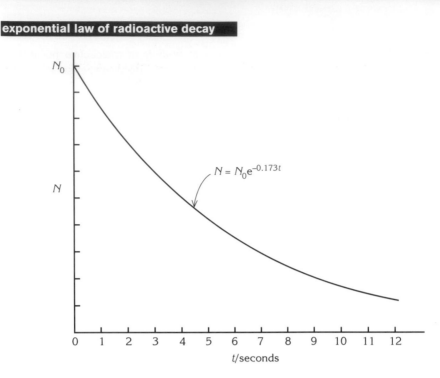

The curve is smooth and mathematical. A magnified section of the graph would contrast the erratic nature of the actual decay with the regularity of the exponential decay curve.

The half-life $t_{1/2}$ of a radioactive nuclide is the value of t for which N equals one half of N_0. The equation becomes:

$$\frac{N}{N_0} = \frac{1}{2} = e^{-\lambda t_{1/2}}$$

$$2 = e^{\lambda t_{1/2}}$$

$$\ln 2 = \lambda t_{1/2}$$

$$= 0.693$$

$$t_{1/2} = \frac{0.693}{\lambda}$$

The only way the theoretical curve can be drawn in practice is by measuring half-life and using it to calculate successive values of N. But half-life is another mathematical abstraction. We get round this by defining half-life as follows:

'The half-life of a radioactive nuclide is the average time taken for half the nuclei in a sample of the nuclide to decay.'

By insisting that half-life is the average value of the measurements there is no theoretical limit to its accuracy. This bridges the gap between $\Delta N /\Delta t$ and dN /dt.

Worked examples:

1. The count rate from an irradiated indium foil falls from 8000 counts per minute to 500 counts per minute in 220 minutes. Find the half-life of the decaying isotope.

   The fractional drop in the count rate in 220 minutes is:

   $$\frac{500}{8000} = \frac{1}{16} = \left(\frac{1}{2}\right)^4$$

   Since 220 minutes is 4 half-lives, the half-life is 55 minutes.

2. The half-life of ^{226}Ra is 1620 years. Calculate (a) the disintegration constant, (b) how long it takes for 60% of a given sample to decay and (c) the activity of 1 g of pure ^{226}Ra. (Avogadro constant L = 6.02×10^{23} mol^{-1}.)

   (a) One year is 31.6×10^6 seconds:

   $$\lambda = \frac{0.693}{t_{1/2}} = \frac{0.693}{(1620 \times 31.6 \times 10^6 \text{ seconds})} = 1.35 \times 10^{-11} \text{ s}^{-1}$$

   (b) When 60% of the radon has decayed, 40% is left:

   $$\frac{N}{N_0} = 0.40 = e^{-\lambda t}$$

   $$\ln 0.40 = -0.916 = -\lambda t$$

   $$t = \frac{0.916}{(1.35 \times 10^{-11} \text{ s}^{-1})} = 6.79 \times 10^{10} \text{ s}$$

   $$= 2.1 \times 10^3 \text{ years} \quad \text{(to 2 s.f.)}$$

   (c) N_0 for 1 g of pure radon is $(6.02 \times 10^{23} \text{ mol}^{-1})/226$. We write:

   $$\text{activity of 1 g of radon} = \frac{dN}{dt} = -\lambda N_0$$

   $$= \frac{(1.35 \times 10^{-11} \text{ s}^{-1})(6.02 \times 10^{23} \text{ mol}^{-1})}{226 \text{ g mol}^{-1}}$$

   $$= 3.6 \times 10^{10} \text{ Bq} \quad \text{(to 2 s.f.)}$$

extra-nuclear electrons: electrons with orbits that take up a large fraction of the atom's volume and whose number equals the number of protons in the nucleus. This is the atomic number Z. The extra-nuclear electrons decide the chemical properties of an atom and most of the physical properties of the materials of which they form a part.

eye: the organ of sight. The main parts of the human eye are shown in the diagram. The image of a distant object is focused on the retina mostly by the curved transparent surface at the front of the eye, the cornea. The lens acts as a fine adjustment for this focusing power; its focal length is controlled by the ciliary muscles. The lens grows stiffer with age and there comes a time when it is no longer supple enough to bring the focal plane of an object at reading distance from behind the eye onto the retina. The subject becomes long-sighted and needs glasses. Others need glasses from birth, either because the curvature of the cornea is a little high for the eye's length (short sighted) or a little too low (long sighted). Short sight is corrected with a diverging lens to push the image back, long sight is corrected with a converging lens to bring the image forward.

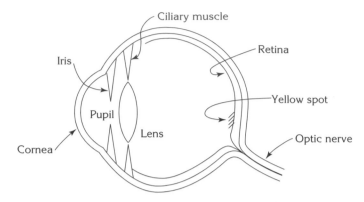

The iris (the coloured part of the eye) controls the size of the hole at its centre, the pupil. The pupil acts as a *diffraction* element and ensures that the image of a bright point object, e.g. a star, forms an image on the retina a few microns across (one micron = 10^{-6} m). This fixes the optimum and actual size of the light sensing cells of the retina, the rods and cones, and limits the detail the eye can resolve. The eye can accommodate an enormous range of intensities and the contraction of the pupil plays no more than a marginal role. The key mechanism involves contractions within the rods and cones.

The cells which are sensitive to light or dark, the rods, are spread across the whole area of the retina. The number of rods in the retina is in the order of 10^9 and, to avoid having an equal number of nerve cells running down the optic nerve, the first stages in image processing take place within the retinal complex. The colour sensitive cells, the cones, are mostly confined to a small patch (the yellow spot) on the retina and opposite the pupil. Our sense of colour across the whole field of view is mostly down to memory. Rods are much more sensitive in low light levels than cones.

If light intensity is low the rods are more active than the cones. Colour is difficult to discern, but any movement is still quite noticeable.

It is interesting how different theories of light are employed to explain different functions of the eye. Image formation on the retina and the design of spectacles fall within the area of geometrical optics. To understand the resolving power of the eye, its ability to distinguish fine detail, we have to appeal to the wave theory of light. Finally, the act of light detection is a chemical process and we need to think of a light beam as a stream of individual photons.

Turning to a different topic, compare eyes with leaves on trees. The chloroplasts in the leaves are spread out over a large area just like rods in the retina. The leaves turn to face the sun just as eyes turn towards whatever we are looking at. Finally, photosynthesis is a chemical process which involves individual photons just like image detection in the eye.

farad: the SI unit of electrical capacitance, symbol F.

Faraday's law of electromagnetic induction: see *electromagnetic induction (laws of)*.

fatigue is the failure of a metal component, its *fracture* under the influence of excessive working. It begins with the initiation of a crack which then propagates through the material under the influence of a fluctuating or alternating *stress*. Provided that the stress reversals are frequent enough and continue for long enough, fatigue occurs at stress levels below the *yield point*.

ferrites are ferromagnetic, multicrystalline materials in which grain size is smaller than domain size. This allows for high permeability and low *hysteresis* loss coupled with short magnetic reversal times. Ferrite cores work at high frequency. See *ferromagnetism*.

ferromagnetism is the powerful magnetic behaviour associated with substances such as iron below their *Curie temperatures*. The origin of ferromagnetism is the capacity of the molecules of some solids to collect together in small magnetic domains. The material is in the unmagnetised state when the domains are randomly orientated and in the magnetised state when magnetic domains are aligned. See *domain theory*.

Feynman diagrams display events which involve subatomic particles. Simplified diagrams list the ingoing particles at the bottom, the outgoing particles at the top and the exchange particles; they are not trajectories or space–time diagrams. The diagrams prepare the ground for the application of the conservation laws to the interaction. The first diagram is particularly important; it contrasts β^- decay (on all syllabuses) with β^+ decay. Both cases involve the decay of a massive but short-lived exchange particle The relevant conservation laws here are baryon number (B), lepton number (L) and charge (Q):

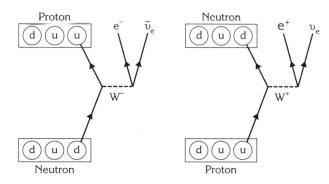

The second diagram shows examples of weak, strong and electromagnetic interactions. All obey conservation laws. The exchange particles belong to the type of interaction.

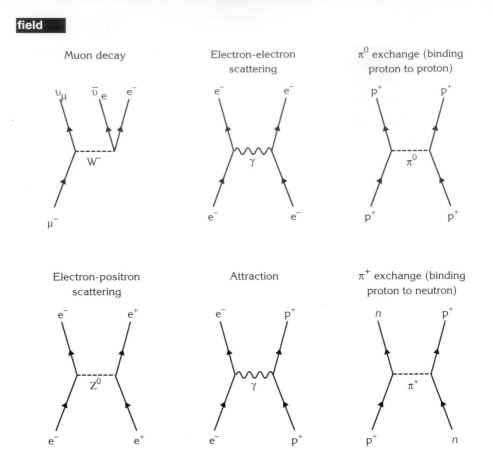

Muon decay

Electron-electron scattering

π^0 exchange (binding proton to proton)

Electron-positron scattering

Attraction

π^+ exchange (binding proton to neutron)

The weak interactions (they must be weak interactions if neutrinos are involved) are muon decay and electron/positron scattering. The electromagnetic interactions show repulsion between two electrons and attraction between an electron and a proton. The strong interactions show the attractions between two protons and between a proton and a neutron. The latter involves a positively charged gauge boson, the π^+ meson.

field is a three-dimensional space which can be mapped with the values of a physical variable. A temperature distribution constitutes a field, so do the velocity vectors along the streamlines of a flowing fluid. But the fields which matter in A level physics are *gravitational*, *electrical* and *magnetic fields*. Each of these quantities is linked to a force field and to a potential field. But magnetic potentials are complex affairs and are omitted. The *force–separation curve* and the *energy–separation curve* are representations of the force and energy fields between the atoms of a solid. (See *field lines (electric)*, *field lines (magnetic)*.)

field lines (electric) are a visual presentation of electric field strength within a space. To give the magnitude and direction of the electric field strength E at any point in the field, the lines must be set up in a particular way. They have the following features:

- the direction of a line at a point is the same as the direction of the electric field vector at that point
- lines begin at a positive charge and end on a negative charge

- the concentration of the lines (the number passing normally through unit area at a point) equals the *electric field strength*

- the lines cannot cross (if they did, the electric field strength would have two directions at the point where they crossed)

- the lines repel one another

- the lines are in a state of tension.

The number of lines leaving unit charge is fixed at a value which ensures that the concentration equals electric field strength. The electric field strength E a distance r from a charge Q is given by the equation:

$$E = \frac{1}{4\pi\varepsilon_0} \cdot \frac{Q}{r^2}$$

This electric field strength is uniform in value across a sphere of surface area $4\pi r^2$. If N is the number of lines leaving charge Q, then:

$$E = \frac{N}{4\pi r^2} = \frac{1}{4\pi\varepsilon_0} \cdot \frac{Q}{r^2}$$

or

$$N = \frac{Q}{\varepsilon_0}$$

The number of lines of force leaving positive charge is $1/\varepsilon_0$ lines per coulomb of charge. See *radial electric field, uniform electric field*.

field lines (magnetic) illustrate the shape of a magnetic field; they show where the field is strong, where it is weak and what direction it takes. To do this, lines of force must have a number of characteristics:

- their direction is from a north-seeking pole to a south-seeking pole

- their density at a point equals the *magnetic field strength* at that point

- they are in a state of tension

- they repel one another.

The diagrams on page 102 show how to use lines of magnetic force to describe some familiar magnetic fields. Notice that both the conductor lines and field lines are broken where they pass behind one another. If these gaps are omitted, the diagrams become ambiguous and worthless. Always include them in your diagrams. See *uniform magnetic field*.

filament lamp: a lamp in which the light source is a wire, heated to a high temperature by an electric power source. The *first law of thermodynamics*, applied to the filament, states that:

$$\Delta U = \Delta Q + \Delta W$$

where ΔU is the increase of internal energy brought about by an influx of disordered energy ΔQ and work done ΔW on the filament.

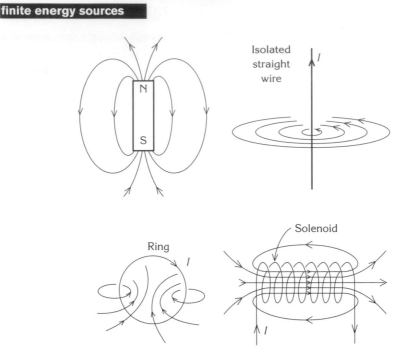

Field lines (magnetic)

There are two cases which matter. When the lamp is first switched on, the temperature is the same as the surroundings and ΔQ is zero. ΔU equals ΔW. ΔW equals $IV\Delta t$ and is positive. So ΔU is positive and the temperature of the filament rises. The second case is for the steady state situation when the filament temperature is constant. ΔU is zero when the temperature is constant. ΔW is a bit smaller because of the increase in resistance and the fall in the current at high temperature, but it is still positive. ΔQ must therefore be negative; it represents the energy radiated from the filament. Only a small proportion of this radiant energy is in the visible band.

finite energy sources are sources of *primary fuels* such as coal, oil or gas, which can be exhausted. (See *energy sources*, *renewable energy sources*.)

fissile materials are minerals which contain a proportion of those chemical elements such as uranium which can release energy in a controlled nuclear fission process. That these materials are also radioactive is incidental. The usefulness of uranium in a nuclear reactor hinges on what happens when it is irradiated with neutrons, not on its very slow radioactive decay. (See *energy sources*, *thermal fission reactor*.)

fission means splitting. The word is used in physics to describe the splitting of a nucleus into two nuclei of roughly equal size and with the emission of much energy. The following is an example of a fission reaction:

$$^{235}_{92}U + {}^{1}_{0}n \longrightarrow {}^{236}_{92}U \longrightarrow {}^{140}_{54}Xe + {}^{94}_{38}Sr + 2\,{}^{1}_{0}n + \gamma + 200 \text{ MeV}$$

fixed point: a temperature at which some recognisable physical event happens. One important example is the *triple point*. Another is the steam point: the temperature at which pure water and steam would be in thermal equilibrium at standard atmospheric pressure.

flavour of quarks: see *quarks*.

Fleming's left hand rule gives the direction of the force on a current-carrying conductor in a magnetic field. It states that if the thumb and first two fingers of the left hand are held at right angles to each other, and if the first finger points in the direction of the magnetic field while the second finger points in the direction of the current in a conductor immersed in the magnetic field, then the thumb will be pointing in the direction of the force on the conductor, as in the diagram.

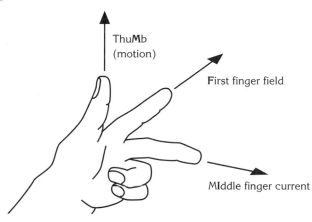

The force is a vector quantity. Fleming's left hand rule gives its direction. The magnitude of the force is given by the equation:

$$F = BIl\sin\theta$$

where F is the force on a conductor of length l carrying current I at an angle θ to a uniform magnetic field of strength B. Notice that $Il\sin\theta$ is the component of the current element in a direction at right angles to B. If θ is a right angle, the equation reduces to the familiar form $F = BIl$. (See *motor effect*, *force on current*, *force on moving charge*.)

Fleming's right hand rule gives the direction of the induced e.m.f. in a conductor. It states that if the first two fingers and the thumb of the right hand are held at right angles to one another, if the first finger points in the direction of the magnetic field and the thumb points in the direction of the motion of the conductor, then the second finger will be pointing in the direction of the induced e.m.f.; the diagram is below.

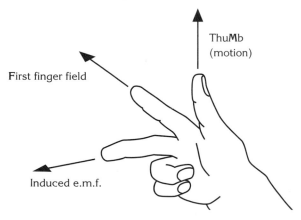

(See *electromagnetic induction*, *laws of electromagnetic induction*.)

flexibility is the capacity of a material to bend easily without breaking and to flip back when the bending stress is removed. The flexibility of a given object is as much a function of its shape as of the material it is made from. For this reason, flexibility is not defined on a scale of values. If two materials are formed into the same shape, the more flexible material will have the lower value of *Young's modulus* to allow it to bend, and a higher tensile strength to stop it from cracking.

flotation (law of): this states that the weight of a floating body equals the weight of fluid displaced. (See *Archimedes' principle*.)

fluid matter is matter which is in the liquid or the gaseous state but not in the solid state. Fluids have no shape of their own. In *solids* the molecules are attached to one another in a rigid structure; forces are transmitted along these structures without being deflected sideways. In fluids the molecules are free to move around; push a liquid surface inwards and the liquid mass further down will move out of the way. (See also *gases*, *liquids*.)

An immediate consequence of this is that we use forces when working with solids and pressures with fluids. Solids transmit forces, fluids transmit pressures.

flux linkage is the product of the flux ϕ passing through a coil and the number of turns in the coil. It is a curious term. It involves thinking of lines of magnetic flux looping from where they start to where they finish. If the magnetic flux loop penetrates a completed loop in a conductor, it creates a link between the circuit and the magnetic flux. If ϕ is the amount of flux, then ϕ is the linkage per conducting loop. If there are N loops in the coil, the total linkage is $N\phi$, which is a measure of the capacity for interplay between magnetic flux and coil. (See *electromagnetic induction*, *electromagnetic induction (laws of)*.

focal length is the distance of the focal point from a lens or spherical mirror. The focal point is the point through which incident light, perpendicular to the plane of the lens or mirror, is directed by the action of the lens or mirror. A lens has two focal points at equal distances from the lens and on opposite sides. Concave lenses and convex mirrors have virtual focal lengths, that is, the light is deflected away from the focus and doesn't actually pass through it. They are mostly omitted from specifications, despite driving mirrors being convex, and short sighted people wearing concave spectacles. The diagram illustrates a convex lens and a concave mirror. See *lens equation*.

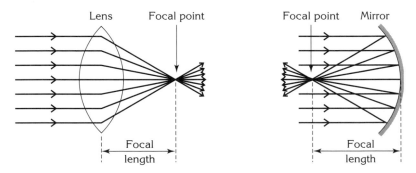

force, in classical Newtonian physics, is a *localised vector* quantity associated with rate of change of momentum; either change of magnitude or change of direction. The SI unit is the newton (N) and the symbol is F. The second of *Newton's laws of motion* defines force in magnitude and in direction. The same law, by implication, defines the newton.

Word equation definitions of force (for constant mass) and the newton are:

force = rate of change of momentum = mass × acceleration
newton = kilogram × metre per second squared (kg m s⁻²)

Forces do not exist in the abstract. A force is the push or pull of body A on body B. We learn from *Newton's third law of motion* that forces always occur in pairs (that is, body B will also push or pull body A – giving two pushes or two pulls). They are usually gravitational or elec-tromagnetic but, for nuclear particle physics, they may come from either the strong or weak nuclear interactions. Contact forces are mostly electromagnetic in origin and sometimes gravitational. They are resolved into normal and tangential components. Tangential compo-nents are usually frictional or viscous forces. The idea of a contact force is introduced in order to facilitate analysis without comment on the origin of the force.

(See also *free-body force diagram, force: a localised vector, kinds of force.*)

force: a localised vector is an expression which reminds us that force is a vector with three characteristics: magnitude, direction and line of action. Forces act along lines and through points. Kick a football through the centre of mass and the football will just fly off; kick it through a point to one side of the centre of mass and it will fly off a little slower than before and it will be spinning. All your work on turning forces underlines the importance of line of action.

force–extension curves are graphs which show how the extension of a material in the form of a wire or a thread varies with the load placed upon it. The information which can be written into a force–extension graph is in many ways better placed on a stress–strain graph. In particular, the magnitudes on the two axes of a *stress-strain* graph depend only on the material and are independent of the size of the specimen. A useful property of the force–extension graph relates to the area between the curve and the extension axis. The area of the shaded strip on the force–extension graph shown in the diagram equals the average force F' for the strip, multiplied by the increase in extension XY. But this product is the force–distance product for this small additional extension (the work done increasing the extension). What is true of the narrow strip is true of the whole area OAB which is no more than a set of many thin parallel strips. So the work done stretching the wire, the elastic potential energy stored in the wire, equals the area OAB, equal to ½F Δx where F is the applied force and Δx is the final extension. For this particular graph, the elastic potential energy stored is 0.15 J.

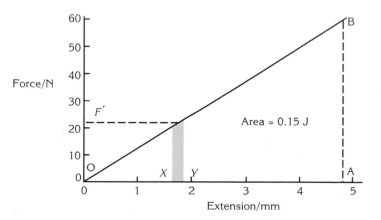

force–separation curve: a graph which shows how the force between two atoms varies with the distance between them. A typical example is shown in the diagram below. The force curve is the resultant of two other curves, a short range but very strong repulsive force and a longer range attractive force. The repulsion wins at small separations, the attraction is dominant when the atoms are further apart. The equilibrium distance between the atoms is when the two forces are equal and opposite. At this point, where the separation is r_0, the curve is approximately straight and this explains why materials obey *Hooke's law*. The gradient at separation r_0 is proportional to the *Young's modulus* for the material and the depth of the deepest part of the curve is proportional to the strength of the material.

The curve has been drawn with repulsion increasing upwards and attraction increasing downwards; Some books prefer it the other way round. As with any force–displacement graph, the area between the curve and the separation axis represents work done. In this instance the energy is calculated from infinite distance inwards, and starts off at zero for infinite distance. The work done is increasingly negative until the point r_0 is reached (r_0 marks the minimum for the *energy–separation* curve) and then it begins to rise again.

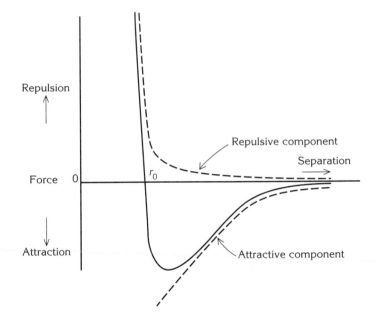

force on current: see *motor effect, Fleming's left hand rule*.

force on moving charge: a phrase which relates to the force on a free charge, usually an electron, moving through a magnetic field at an angle to the field. The force on an electron in an electric field equals Ee whether or not the electron is moving.

The force F on charge q moving through a uniform magnetic field of strength B with velocity v along a path at right angles to the field direction is given by the formula:

$$F = Bqv$$

The direction of the force is given by *Fleming's left hand rule*. Remember when applying the rule that the current direction is opposite to the direction of motion for negative charge.

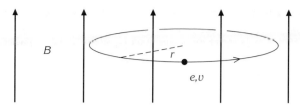

The diagram illustrates an important special case. A beam of electrons, all travelling with speed v at right angles to an extensive uniform magnetic field of strength B is pulled by the force F into a circular path of radius r. The centripetal force equals Bev.

$$Bev = \frac{mv^2}{r}$$

$$r = \frac{mv}{Be}$$

(See also *electron beams*.)

forced oscillations are the oscillations of a damped system acted on by a periodic driving force. Were the system not damped, the oscillations would soon go wildly out of control. If the system is damped, an equilibrium situation is reached when per cycle, the energy transferred to the system from the driver equals the energy transferred from the system through the damping mechanism. The amplitude of this equilibrium state for a given driver amplitude depends on the frequency of the driver. It is a maximum when the driver frequency is the same as the natural frequency f_0 of the undamped system. This is called *resonance.*

The link between resonance and physical damage is connected to this maximum amplitude in an obvious manner: the vibrating system is subject to maximum distortion. Less obvious, but just as important, is the fact that the rate of energy flow from the driver, through the system into the damping mechanism is also at its maximum value. This is the principle behind tuning circuits, musical instrument design and many other engineering applications.

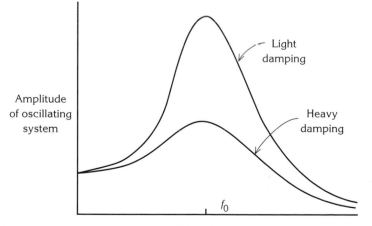

forces in equilibrium are forces which cancel one another out in two respects: they provide no net force and no net couple. A stationary object acted upon by several forces in equilibrium, will neither move nor turn round. At A level, the work is always confined to a set of forces which all operate in the same plane, (coplanar set).

The two conditions for a set of coplanar forces to be in equilibrium (no net force, no net couple) can be stated in two ways:

First way:

1   the sum of the components of all the forces in any one direction is zero

2   the sum of the clockwise moments of the forces about any point in the plane equals the sum of the anticlockwise moments about the same point.

Second way:

1   the vectors representing the forces, if placed end to end, trace out a closed polygon. (If there are three forces, this statement becomes the *triangle of forces* statement)

2   the lines of action of all the forces meet at a point.

These ideas are more part of applied mathematics than of physics and would only be tested in a superficial way in a physics examination.

fossil fuels, such as coal, oil and gas, are derived from once-living matter. Peat is mostly described as a fossil fuel but sometimes listed as a biofuel. The trouble is that some peat isn't very old. (Avoid mentioning peat if possible.) (See *energy sources*.)

four fundamental interactions account collectively for the properties of the fundamental particles, the subatomic particles and of atoms and matter in bulk. They are the strong interaction, the electromagnetic force, the weak interaction and the gravitational force. Their relative magnitudes at a given distance are in the proportions $1 : \frac{1}{137} : 10^{-5} : 10^{-39}$.

Electromagnetic and gravitational forces apply to everyday objects and their actions are usually handled algebraically with inverse squares law equations. At the sub-atomic level all sense of position and timing is lost and we have to think more of events than of processes. Subatomic events are described in *Feynman diagrams*. These diagrams identify the relevant particles before and after the event and identify the responsible interaction by showing the relevant *exchange particle*.

The gravitational interaction. The gravitational interaction is so weak that it might not be expected to operate at all. But there are two points in its favour. First, it operates across large distances which rule out the strong and weak interactions . Secondly, all lumps of matter of any appreciable size are electrically neutral and this rules out the electromagnetic interaction. On the large scale it is the only interaction left. The exchange particle, the graviton, has yet to be detected.

The electromagnetic interaction. The electromagnetic interaction is effective between charged particles, that is, all quarks, some leptons and some sub-atomic particles. It is more than a hundred times weaker than the strong interaction but it takes over outside the nucleus and is the principal force accounting for atomic stability and structure, molecular structure and the chemical and mechanical behaviour of matter in bulk. The gravitational force takes over at larger distances because matter in bulk is electrically neutral. The exchange particle, a virtual photon, transmits the interaction between two charged particles.

The strong interaction. Two protons in a nucleus each carry charge e$^+$ and may be as close together as 10^{-15} m. Coulomb's law, applied to this situation, gives a force of repulsion of about 230 N. Unless there were some other, much stronger force in action within the nucleus, electrostatic repulsion would tear the nucleus apart. The interaction which counters this destructive force was called the strong interaction and is now associated primarily with quarks. This strong interaction is shown by *alpha particle scattering* to be a short range force which does not act over distances beyond about 10^{-15} m. This is a distance roughly equal to proton or neutron diameter. This range is far enough to hold a nucleus together but is ineffective at greater distances. The word 'strong' carries a meaning closer to that in a 'strong glue' which holds two pieces of wood together once contact has been made, not an attraction that pulls them together from any great distance.

Sub-atomic particles which are built from quarks and which are consequently affected by the strong interaction are called *hadron*s. Charged hadrons are subject to all four interactions, neutral hadrons to three of them. The strong interaction holds quarks together in triplets to form baryons or in pairs to form mesons. Nuclear reactions are the typical events in which the strong interaction operates.

The weak interaction. The weak interaction is best known as the force which operates during *beta decay*. It is a complex force which is treated together with the electromagnetic force as the electroweak interaction. It is about a thousand times weaker than the electromagnetic interaction and operates only within distances of about 10^{-17} m. If it was weak in the ordinary sense of the word, it would be swamped. But the word 'weak' implies a low probability of its coming into play. This probability increases, that is the interaction becomes more probable, as the energies of the participating particles increase. Apart from beta decay, it comes into its own at about 100 GeV. The exchange particles for the weak interaction are labelled W$^+$, W$^-$ and Z^0 particles.

The weak interaction is the only one of the four interactions to which neutrinos are susceptible, so any event involving a neutrino is controlled by the weak interaction. Energy, mass and charge are interchanged between leptons, quarks and their antiparticles by the electroweak interaction. This is therefore the interaction which controls hadron and quark decay and transformation events. Such events are always in the direction of lower mass and this explains why most of the mass in the universe is wrapped up in electrons, protons, neutrons and electron neutrinos. Powerful fluxes of electron neutrinos are emitted continuously from the Sun and other stars as a consequence of the hydrogen fusion reaction. See *exchange particle, Feynman diagram, conservation laws, uncertainty principle.*

fracture is a breaking of a solid mass by the growth of a small crack. The origin of a fracture in a metal is likely to be *fatigue*. In a ceramic it is likely to be the spreading of a crack through the hard, unyielding material.

free-body force diagram: a way of listing the forces which act on a body with a view to determining the resultant force. Never draw more than one body. Do not draw the ground, water or anything else the body is resting on, just a labelled blob to represent the body on which the forces are acting. (A number of examples are shown in the diagram on page 110.)

A resultant force (e.g. net downward force, centripetal force) should not appear on the diagram; only the individual forces such as weight, upthrust or drag should be marked and each of these must be labelled. If you want to include the direction of a velocity, say, label it clearly and put it separately, well to the side of the diagram and separated from it.

Sometimes you have to draw two forces upwards or downwards, both passing through the centre of mass. Put these side-by-side, close to the correct spot and label them clearly.

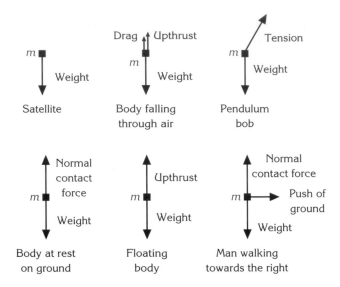

free fall: the motion of a body whose weight is the only force acting on it. See *acceleration of free fall, weightlessness.*

frequency is the number of times per second an event occurs. The symbol for frequency is *f* and the unit of frequency is 'events per second'. For example, the frequency of cars arriving at traffic lights is in 'cars per second'; for spectators through a turnstile, it is 'spectators per second'.

For simple harmonic motion and wave motion the event which matters is a cycle and frequency is the number of cycles per second. The SI unit for frequency in cycles per second is hertz (Hz). (Note, hertz does not mean second^{-1}.)

The algebra for simple harmonic motion and alternating current theory is often simplified by using the symbol ω for $2\pi f$. ω is sometimes called angular frequency, sometimes angular speed and sometimes pulsatance but mostly it is simply called omega. The relationship between ω and *f* is set out in the following equation:

$$\omega = 2\pi f$$

ω in radians per second $= 2\pi$ in radians per cycle $\times$ *f* in cycles per second

$$\frac{radian}{second} = \frac{radian}{cycle} \times \frac{cycle}{second}$$

There is no difference in principle between this last equation and an equation converting a mass measured in pounds to the same mass measured in kilograms:

4 pounds $=$ 2.205 pounds per kilogram $\times$ 1.814 kilogram

Frequency measured in cycles per second can be converted into angular frequency in radians per second; 2π radians per cycle is the conversion factor.

Another example of a frequency is activity in radioactivity; it is the number of disintegrations per second. The unit for 'disintegrations per second' is the becquerel (Bq). You will not confuse becquerel with hertz; don't confuse either with second^{-1}. Count rate is a frequency measured in counts per second.

One last point. The ω used in AC theory is strictly the *angular speed* (in radians per second still) of the phasor used on the phasor diagrams. This is why ω is sometimes called angular speed. The term ωt will be a number of radians, whence $\sin \omega t$, $\cos \omega t$ and $\sin 2\pi ft$, $\cos 2\pi ft$.

frequency division multiplexing (FDM) is a technique which enables the use of a single telephone wire to carry several sound messages simultaneously and in analogue form. The term FDM is a little forbidding at the first meeting, but it means no more than it says – different frequency ranges in a signal are used to do different jobs. It is about the economic use of resources. A high frequency carrier wave (several megahertz) has a spread of frequencies called its bandwidth of perhaps 20 kHz. This bandwidth is split up into half a dozen bands, each of which is wide enough to carry a sound-frequency wave without distortion. To the amplitudes of each of these elements of the bandwidth is added the amplitudes of one of the sound-frequency waves. This addition action feeds the carrier wave with both the amplitude and the frequency of the original sound wave. The carrier wave can, then carry several telephone messages at once. A demultiplexer separates the different telephone messages carried on the different frequency bands after reception. The comparatively high frequency range of visible light waves means that the capacity of optical fibres to carry simultaneous telephone messages is proportionately much larger. They carry many thousands of transmissions simultaneously. *Time division multiplexing* (TDM) is used for the transmission of information in digital form, as in *pulse code modulation* (PCM).

frequency modulation (FM) means transmitting information by means of a carrier wave whose amplitude is kept constant but whose frequency is changed in proportion to the amplitude of the information wave (e.g. a radio wave). Since these frequency changes keep pace with the amplitude changes, the frequency of the information wave is transmitted too. See *modulation*.

friction is a contact force between two, usually solid, surfaces which acts parallel to the surfaces either to prevent their relative motion, or oppose it without entirely preventing it. In the latter case some driving force must work against friction, energy will be transferred and the surfaces will warm up.

It is instructive to think about what exactly happens when we walk across a flat piece of ground. Start first with what happens when we jump up into the air. The body pushes down hard on the ground to push the Earth away. The normal reaction at the ground increases and pushes us up. We can only be pushed upwards by an upwards force from outside ourselves and this is the increased normal reaction from the ground. Equally, when we walk we push backwards and the frictional force from the ground pushes us forward. And what about the work done, where does the energy come from? We have to remember here that the human body is not rigid. The leg muscles do work in pushing the rest of the body away from the foot which is touching the ground, they do work in transferring kinetic energy to part of the body. The body is using the feet to push the Earth backwards. The external force which pushes the body along the ground is friction while the energy source is our muscles.

fuel rods are the long metal-clad containers, filled with nuclear fuel, which are inserted into the *moderator* assembly of a nuclear power plant.

full-wave bridge rectifier: see *rectification, half-wave and full-wave*.

fundamental law of radioactive decay states that the activity, $\Delta N / \Delta t$ of a radioactive nuclide is proportional to the number, N, of undecayed atoms of the nuclide in the sample. That is:

$$\frac{\Delta N}{\Delta t} = -\lambda N$$

where λ is the *decay constant* for the nuclide and the minus sign indicates that ΔN is a drop in the value of N. See *exponential law of radioactive decay*.

fundamental particles are those subatomic particles which are thought not to have parts and which form the building blocks for other sub-atomic particles. They are too small to examine for any possible structure. Many of them are unstable and quickly decay into other fundamental particles. There are two sets of fundamental particles: *leptons* and *quarks* and their typical diameter is smaller than 10^{-18} m. *Exchange particles* are virtual particles which can only exist for a short timespan during an interaction. See *four fundamental interactions*, *Feynman diagrams*.

fusion means joining together. The word is used in physics to describe the building up of larger nuclei from smaller nuclei with the emission of energy. (See *binding energy*, *fission*.)

What other subjects are you studying?

A–Zs cover 18 different subjects. See the inside back cover for a list of all the titles in the series and how to order.

galaxies: The concept of a galactic universe properly began with William Herschel. Herschel assumed that the faintness of a star is an indicator of distance. By counting and recording the brightnesses of as many stars as he could see, he showed that stars were distributed over a disc-like region of space.

A galaxy is now recognised as a group of about 10^{11} stars and spread across about 10^5 light years. There are considerable differences in size, shape and stellar distribution. Galaxies are about 10^6 light years apart (i.e. 10 galactic diameters), the farthest is about 10^{11} light years away and there are about 10^{13} of them. See *Hubble's law*.

gamma camera: a device for generating an image of an extended gamma radiation source. The essential features are shown in the diagram. The source, typically a radioactive tracer isotope such as ^{99m}Tc, is attached to a carrier compound which takes it to the organ under test. Gamma radiation which gets through a lead collimator reaches a sodium iodide scintillator crystal and is absorbed. A proportion of the absorbed energy re-appears in the form of visible photons. The different readings from the bank of photomultipliers enables the precise position of the gamma emission event to be calculated. The computer uses the information from tens of thousands of such events to build up an accurate picture of the gamma source, from which an image of the organ can be displayed on a screen.

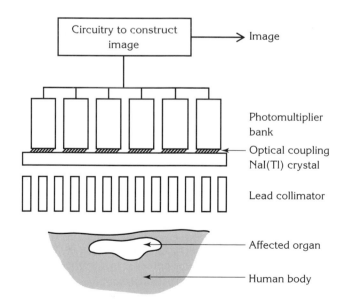

gamma emission is the release of high energy, very short wavelength photons (*gamma radiation*) from an excited nucleus. There is no change in the proton number or the atomic number for the nucleus: it is not a type of radioactive decay.

gamma radiation is very short wavelength *electromagnetic* radiation emitted from nuclei which are in heightened energy states. A nucleus may be in a heightened energy state because it is the radioactive decay product of a parent nucleus, it may have been produced in a fission process or it may have been struck by a high-energy particle such as a neutron in the shielding of a nuclear reactor. There is no change in the structure (decay) of the nucleus when the gamma radiation *photon* is emitted, only a re-arrangement and an energy relaxation.

Gamma radiation photon energies are mostly within the range 1.2 keV (1.9×10^{-16} J) to 1.2 GeV (1.9×10^{-10} J). These energy values give wavelengths from about 1 nm (1×10^{-9} m) to about 1 fm (1×10^{-15} m). The corresponding frequency values are from about 0.3 EHz (0.3×10^{18} Hz) up to about 0.3 YHz (0.3×10^{24} Hz). Since there is no known upper energy limit, wavelength values vary somewhat from one book to the next. One reason for the stress on energies is that photon energies can be measured. Wavelengths are rarely measured, and then only if there is a need to verify that particles are photons and not something else. The lower energy end, up to a few MeV (a few per cent of the overall energy range), overlaps with the *X-ray* segment, but X-rays do not come from nuclear relaxation processes.

The properties of gamma radiation are what you would expect of high-energy, short-wavelength particles. Their paths are effectively straight with little evidence of diffraction and their interaction with matter is largely destructive. The straightness of their paths ensures the application of the *inverse square law* to the fall away of beam intensity with distance from the source. Low energy gamma radiation is strongly scattered. At higher energies, photoelectric absorption becomes more likely. As photon energy increases still further, Compton scattering begins to take over and finally, pair production. Short wavelengths mean that interaction probabilities are low and this favours deep penetration of matter by gamma radiation before destructive absorption begins. From a biological or medical point of view, this allows deep body damage so gamma radiation sources should be efficiently shielded and kept far away.

Detection of gamma radiation is largely by photography or scintillation counter. GM tubes are largely transparent to gamma radiation and only a small proportion of an incident beam is detected. Choice of method depends on the seriousness of the investigation. Photography is mostly limited to low level monitoring work, such as lapel badges. *GM tubes* work indirectly, they measure the X-rays emitted when gamma radiation reacts with the material of the tube. The output correlates with gamma radiation beam intensity but precision is poor unless there is close calibration of the tube against the incident gamma radiation wavelength pattern. *Scintillation counters* work best but they are expensive. The detector is generally a single crystal (a 4 cm cube) of sodium iodide laced with thallium. A photomultiplier tube is mounted across one face of the crystal with good optical contact. A single gamma radiation photon enters the crystal and releases all its energy in producing a stream of visible radiation photons which are detected by the photomultiplier arrangement. The instrument records time of arrival of individual gamma radiation photons and their energies. This close monitoring is essential in experiments in nuclear spectroscopy which aim to investigate nuclear structure.

Uses of gamma radiation are linked to their high powers of penetrability. Leaks in gas and water pipes can be detected deep underground, rock structures can be investigated for depth and continuity, metal components can be examined for internal cracks. Medical uses range from the destruction of cancerous tissue to the exploratory powers of the gamma camera. (See also *electromagnetic spectrum, GM tube, radiation detectors*.)

gases are fluids in which molecules are spread as far apart as the container will allow. The average molecular kinetic energy is too high for the intermolecular forces to hold the molecules together; shape and volume are decided by the container.

Gases crop up at a number of points in A level physics. The gas laws are based on experiment and enable us to calculate changes in the relevant physical variables such as pressure, volume or temperature. Pressure, volume and temperature are defined without reference to the gas laws. Kinetic theory offers an explanation for gas law behaviour in terms of atomic theory and the energy of motion. This, in turn, leads on to a better understanding of the thermal properties of matter generally, and of evaporation and change of state in particular. Along an almost separate track, air is an elastic fluid and can carry those oscillations which we recognise as sound waves. Because gases are fluids, they transmit pressure, not force.

Gas pressure and liquid pressure are the same kind of thing, one can balance the other. But their origins are largely different. Liquids have densities about a thousand times higher than gases and the pressure created by gravity at the bottom of a modest depth of liquid is easy to measure. A 76 cm column of mercury balances a column of air of constant density, about 8 km high. The pressure in a container of liquid increases noticeably with depth, but that in a container of gas is almost wholly a consequence of molecular motion and is effectively the same throughout any container of modest size. Gas pressure in a closed vessel increases proportionately with rising temperature at constant volume. (See *ideal gas equation, kinetic theory of gases, solids, liquids*.)

gauge bosons: see *exchange particles*.

gauge pressure is the reading of a pressure gauge which, like a Bourdon gauge, measures the difference between the pressure inside the instrument and atmospheric pressure outside.

general gas equation: see *kinetic theory of gases, ideal gas equation*.

general relativity sets out to extend the space–time concepts of *special relativity* to all reference frames, in particular to accelerating reference frames. An important outcome of the theory was the correct quantitative forecast for the deflection of light by the Sun's gravitational field.

generator: a device for transforming mechanical energy into electrical energy by means of a coil rotating in a magnetic field.

The principle should be clear from the diagram on page 116. A coil constructed from many turns of fine wire is mounted on a spindle in a magnetic field. It is made to rotate on the spindle by a turning force applied to the pulley. The coil rotates in the magnetic field and the e.m.f. induced in the coil is picked up from the two commutator rings. The output voltage is the familiar sine wave alternating voltage the frequency of which is decided by the angular speed of the coil. The magnitude is decided by the strength of the magnetic field, the number of turns in the coil, the area of the coil and (again) the angular speed.

The angular speed of the coil governs the output frequency because output voltage goes through one complete cycle each complete turn of the coil. The angular speed influences the output voltage amplitude because an increasing angular speed causes the magnetic flux linkages to change at a faster rate. Also, if the angular speed is zero, the output voltage is zero. All this should be clear from the equation for the output voltage:

$$V = \omega NAB \sin \omega t = V_0 \sin \omega t = V_0 \sin 2\pi f t$$

where V is the output voltage at time t, N is the number of turns, A is the area of the coil and B is the magnetic field strength of the uniform field in which the coil is mounted.

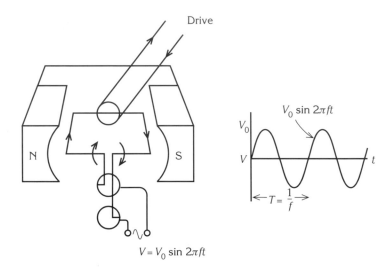

$$V = V_0 \sin 2\pi f t$$

If current is drawn from the output, the generator will be supplying electrical power. The direction of the current in the coil, shown on the diagram, is found from *Fleming's right hand rule*. If we apply Fleming's left hand rule to this current, we find the magnetic field interacts with the rotating coil to give a slowing down turning force. The coil is prevented from slowing down by the mechanical turning force applied to the pulley. In this way, the mechanical work needed to turn the coil is regained as electrical energy in the output circuit. Draw more current and more work must be done on the spindle to keep the generator turning at the right angular speed. (See *electric motor, electromagnetic induction, laws of electromagnetic induction*.)

geothermal energy is energy drawn directly from hot parts of the Earth's interior which happen to be close enough to the surface to use in this way. Suitable sites are rich enough to encourage optimism but they are not very widespread. (See *energy sources*.)

glass fibres are fibres of glass which are too thin to contain any weakening deformations within their structure and which are, therefore, able to support considerable tensile stresses. They can be used collectively as a thermal insulator or, when set in a bonding matrix, to form a light composite material (popularly called 'fibre glass') of great strength which can be moulded as required. (See *optical fibre*.)

glasses are those apparently *solid* substances which when heated slowly soften into the visibly *liquid* state without an identifiable melting point. The reason for the smooth transition from the apparently brittle solid to a recognisably liquid state is that, at the lower temperature, atoms are attached to one another but not in an orderly and regular crystal form. There is no solid state bonding to break down. The strong forces between the atoms accounts for the brittleness and the absence of order accounts for the irregularity of the broken surface.

gluons are the exchange particles associated with the strong interaction and the strong binding of *quarks* within *hadrons* and of *protons* and *neutrons* within a nucleus. There are eight kinds of gluons, which match the colours (not the flavours) of the individual pairs of quarks that they bind together. See *four fundamental interactions*.

gradient is a measure of steepness. It is defined for a straight line graph as the tangent of the angle between the line and the *x*-axis (abscissa). An example is shown in the diagram below. Notice that the gradient is the ratio of two lengths (two sides of a triangle), each measured against the corresponding axis. Remember always:

*   to draw and use as large a triangle as possible

*   to evaluate the length of each of the two lines from the corresponding axis and

*   to give as the unit for the gradient, the ratio of the *y*-axis unit to the *x*-axis unit.

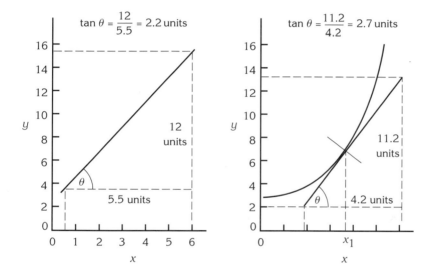

The second graph is not straight, but curved. The instantaneous *rate of change* of the *y* variable with respect to the *x* variable when the value of the *x* variable is x_1 is the gradient of the tangent to the curve at this value of *x*. The gradient is a property of the tangent.

When drawing a tangent to a curve, it is often helpful to draw a short line through the curve at the point of interest and at right angles to the curve, i.e. draw a short 'normal' to the curve. The tangent can then be drawn at right angles to this normal. The sharper the curve, the better it is to follow this procedure.

Avoid the word 'slope'. It means too many things. (See also *limiting value*.)

graph: a way of displaying how one physical or mathematical quantity changes in value should another physical or mathematical quantity change in value. The latter quantity is sometimes called the independent variable; it is usually plotted across the page along the *x*-axis (the abscissa). The other quantity is called the dependent variable and would be plotted up the page along the *y*-axis (the ordinate).

You should draw graphs as large as possible. A graph is an instrument, a small graph is an inaccurate one. Each centimetre along either axis should correspond to 1, 2 or 5 units. 3 units or 7 units makes it impossible to plot something like 5.3 or 15.7. You will be penalised in an examination for using an awkward scale.

The rule that titles should be placed above graphs but below diagrams or figures is often ignored. Try to keep to this rule in laboratory reports. It avoids the title being confused with the labelling of the x - axis.

The axis labels should include any multiple or sub-multiple of the unit so that the numbers on the axis are precisely defined in the label. Sometimes this proves difficult. If a typical x value (the size of a water droplet for example) is, say, 1.2×10^{-5} m then label the axis:

water droplet diameter / 10^{-5} m

The mark 1.2 along the x-axis now means what you want it to mean.

graph (area between line and *x*-axis): this area is significant throughout A level physics. The frequency with which these topics are appearing in examinations is increasing. This is because of a growing emphasis on a particular skill, namely the ability to present information in numerical, algebraic or visual form. The more common examples are in the following list.

Graph [y-axis against x-axis]			Interpretation of area
speed against	time	$\longrightarrow$	distance travelled
force against	distance	$\longrightarrow$	work done by force
force against	time	$\longrightarrow$	impulse applied by force
force on helical spring against	extension	$\longrightarrow$	energy stored in spring
stress applied to wire against	strain	$\longrightarrow$	energy stored per unit volume
gas pressure against	volume	$\longrightarrow$	work done on gas
charging current for capacitor against	time	$\longrightarrow$	charge stored
potential difference across capacitor against	charge stored	$\longrightarrow$	energy stored
potential difference against	charge transferred	$\longrightarrow$	work done
potential difference against	current	$\longrightarrow$	electric power
current against	time	$\longrightarrow$	charge flowing in a given time
radioactive decay rate against	time	$\longrightarrow$	number of nuclei having decayed

Notice that the unit for the area is the product of the units for the two axes.

Questions often expect the candidate to estimate the magnitude of the physical quantity represented by the area. Where the area is irregular, its magnitude is best found by counting the number of small squares that just fit inside it. The height of the small square measured against the y-axis should be multiplied by the width of the small square measured against the x-axis. Multiply this product and its unit by the number of small squares that fit into the relevant area. The final product equals the value of the physical quantity represented by the area.

gravitation is the mutual attraction between any two masses in the universe. (See *Newton's law of gravitation*.)

gravitational constant G is the constant in the mathematical expression of *Newton's law of gravitation*. Its value is 6.67×10^{-11} N m^2 kg^{-2}.

It happens that if the distances of the planets from the Sun and the periodic times of travel of the planets round the sun are measured, Newton's law of universal gravitation will only enable us to calculate the product of the gravitational constant and the Sun's mass. Neither is known separately. To find the mass of the Sun, or the mass of any planet with a moon, we must break the deadlock by measuring the gravitational constant 'G' in the laboratory. The principle of the experiment is as follows:

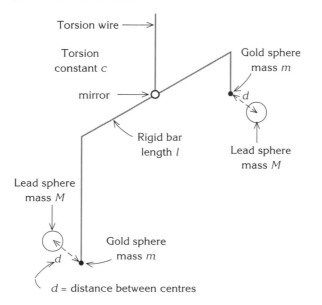

d = distance between centres

The apparatus is assembled without the lead spheres in position. The torsion constant c for the supporting thread is measured (if the thread is twisted through an angle θ radians, then the restoring couple trying to unwind the thread is $c\theta$). The apparatus is allowed to settle down and the direction of the rigid rod is determined optically with some precision. The lead spheres are put in place and the top of the torsion thread turned through an angle, θ, just large enough to keep the rigid supporting rod pointing in its original direction. θ is then measured and G is found from the relation:

$$c\theta = G\frac{mM}{d^2} \times l$$

The masses of the spheres, m and M, the length l of the rigid bar, and the distance d between their centres are measured. The value of G can then be calculated.

Once the value of G is known, the masses of the sun, the planets and a few double stars can all be calculated.

gravitational field strength 'g' at a point in a gravitational field is the force per unit mass acting on a small mass placed at that point in the field. It is a vector quantity and has the unit N kg^{-1}. The word equation definition is:

$$\text{gravitational field strength } g = \frac{\text{gravitational force}}{\text{mass}}$$

$$\text{unit for gravitational field strength } g = \frac{\text{newton}}{\text{kilogram}} = N\,kg^{-1}$$

The difference between gravitational field strength and the acceleration of free fall is linked to the difference between gravitational mass and inertial mass. We can write, using the idea of a gravitational field:

$$F = G\frac{mM}{r^2} = mg \qquad \text{where} \quad g = \frac{GM}{r^2}$$

g is the gravitational field strength a distance r from a point mass M. m is the gravitational mass of the object placed at the relevant point. But we could, instead, follow *Newton's second law of motion* and write:

$$F = mg$$

where F is the force which will act on the object, m is its inertial mass and g is the expected acceleration, called the acceleration of free fall, with the unit m s^{-2}. One achievement of Einstein's relativity theory was to show that these two treatments are essentially identical. (See *acceleration of free fall*.)

gravitational potential (radial fields) at a point in a radial field is the work done per unit mass against the field, in bringing a small mass from infinite distance to the point. Since gravitational fields are attractive and the potential at infinite distance is zero, all points within the field have negative values of potential. Gravitational potential is a scalar quantity with SI unit J kg^{-1}. The symbol used is mostly V but sometimes V_r or $V\{r\}$. A radial gravitational field is one in which the field strength has the same magnitude at all points at a given distance from the centre.

A few lines of calculus (not in the syllabuses) are sufficient to show that the gravitational potential a distance r from a point mass is given by:

$$V = -\frac{Gm}{r}$$

Worked example

Given that the mass of the Earth is 6.0×10^{24} kg and that its radius is 6.4×10^6 m, sketch a diagram of the Earth's gravitational field and label the gravitational potential at the Earth's surface and at distances of 10 000 km, 20 000 km and 30 000 km from the Earth's centre. Calculate the minimum vertical speed at which a rocket must leave the Earth's surface to reach a height of 20 000 km.

At the Earth's surface:

$$V = -\frac{Gm}{r} = -\frac{(6.67 \times 10^{-11}\ N\,m^2\,kg^{-2})(6.0 \times 10^{24}\ kg)}{(6.4 \times 10^6\ m)}$$

$$= -6.2 \times 10^7\ J\,kg^{-1} \quad \text{(to 2 s.f.)}$$

The other three values of V are -4.0×10^7 J kg^{-1}, -2.0×10^7 J kg^{-1} and -1.3×10^7 J kg^{-1}.

To reach the 20 000 km level, the initial kinetic energy of the rocket must equal the gain in potential energy:

$$\tfrac{1}{2}mv^2 = m \times \{(-2.0 \times 10^7\ J\,kg^{-1}) - (-6.23 \times 10^7\ J\,kg^{-1})\}$$

$$= m \times \{4.2 \times 10^7 \text{ J kg}^{-1}\}$$
$$v = \sqrt{(8.4 \times 10^7)} = 9.2 \text{ km s}^{-1} \quad \text{(to 2 s.f.)}$$

gravitational potential (uniform fields) at a point in a uniform field is the work done per unit mass bringing a small mass to that point from some reference point where the gravitational potential is arbitrarily zero. Gravitational potential is a scalar quantity with SI unit J kg^{-1}. The symbol used is V. Gravitational potential enables us to solve many problems by considering energy transformations only. Forces have directions and are more complicated to work with than potentials which do not.

Whether or not a field is uniform is a question of scale. The gravitational field near the Earth's surface is uniform over vertical or horizontal distances which are very small compared with the Earth's radius. Go too far horizontally and the direction of the Earth's gravitational field changes, too far vertically and the magnitude changes. The meaning of 'small' depends on how many significant figures are involved. Over a distance of a few kilometres horizontally or vertically, just above the Earth's surface, the gravitational field can be treated as uniform.

The diagram below shows the essential features of a uniform gravitational field. The magnitudes and directions of the gravitational field strength vectors are the same at all points in the field and the lines of equipotential are parallel to one another, normal to the field lines and equally spaced.

An equipotential surface in a gravitational field is a plane within which a mass can be moved around without exchanging energy with the field. The arbitrary nature of the numerical values of the gravitational potentials does not matter: in uniform fields, it is potential difference that matters.

With a gravitational field strength of 9.8 N kg^{-1}, the work done against the gravitational field when moving a mass of 1 kg a distance of 10 m against the direction of the field (i.e. upwards) is 98 N m kg^{-1} or 98 J kg^{-1}. This is the change in gravitational potential over a 10 0m increase in height and it explains where the equipotential values come from. The height at which V is zero is wholly arbitrary.

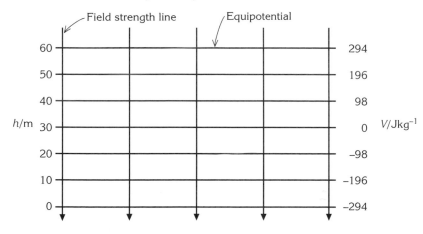

It is well known that change of potential energy equals mgh. The term gh is the gravitational potential V with respect to an arbitrary height where $h = 0$ and $m \times gh$ is the work done moving the mass m through the height h. For a falling body:

$$mgh \; = \; \text{potential energy released} \; = \; \text{kinetic energy gained} \; = \; \tfrac{1}{2}mv^2$$

There are no uniform gravitational fields in nature. The ideas outlined above are tested to see whether or not you understand the concept of gravitational potential and the usefulness of a scalar quantity in solving problems connected with vector fields.

(See *gravitational potential (radial fields)*.)

Do you know we also have A–Zs for:

- Chemistry
- Biology
- Mathematics
- ICT & Computing?

Ask for them in your local bookshop or see the inside back cover for ordering details.

hadrons are subatomic particles which are composed of quarks and hence subject to the strong interaction. Particles which are not hadrons are inert to the strong interaction. Hadrons are subdivided into *baryons* (spin ±3/2, ± 1/2) and *mesons* (spin ±1, 0). Baryon number B, which has value 1 for baryons and 0 for mesons, is conserved in all reactions involving hadrons. Another quantum number associated with hadrons is the strangeness quantum number S. S has integral values and is conserved in nuclear reactions controlled by the strong interaction but not in nuclear reactions controlled by the weak interaction. Charge Q is a further quantum number which must be conserved.

There are so many different kinds of hadrons with such a large range of properties that in order to impose some kind of order on the data, it has been proposed that hadrons are constructed from a more primitive set of particles called *quarks*. Baryons are formed from three quarks; mesons are formed from two quarks. The quarks are held together by the strong interaction; this ensures that hadron diameter is always round about 10^{-15} m, where 10^{-15} m is the maximum distance over which the strong interaction can act.

The quarks in a hadron are held together by the *exchange particles* corresponding to the strong interaction; these exchange particles are called gluons. There are eight different kinds of gluons, this large number being needed to accommodate interactions between quarks of different colour. (See *quarks, baryons, mesons, gluons, four fundamental interactions.*)

half-life of a radioactive nuclide equals the average time needed for half the nuclei in a sample of the nuclide to decay. The symbol used is $t_{1/2}$. It is related to the *decay constant* λ by the relationship

$$t_{1/2} = \frac{\ln 2}{\lambda} = \frac{0.693}{\lambda}$$

There are two curious points about half-life that are worth remembering. First, it has the largest range of values of any known physical quantity. Secondly, there is no known way of changing the half-life of a specimen of a particular nuclide; high temperatures, strong magnetic fields etc have no effect since the extra-nuclear electrons form an efficient shield. (See also *decay constant, randomness and radioactive decay, exponential law of radioactive decay.*)

Hall effect: a difference in potential V_H set up in a conductor at right angles to the current flow direction by a magnetic field B. The diagram on page 124 shows a thin rectangular conductor with the current I flowing parallel to the long side. A magnetic field of strength B is switched on at right angles to the face of the specimen with the largest area. The force on the current in the magnetic field causes it to be deflected downwards; check this with *Fleming's left hand rule*. The concentration of charge on the lower edge sets up a voltage which opposes further deflection of the moving charge. Equilibrium is reached when:

$$V_{H} = \frac{BI}{nqt}$$

where n is the number of charge carriers per unit volume of conductor, q is the charge per carrier and t is the thickness of the specimen in the direction of the magnetic field.

The Hall effect is significant for a number of reasons. It provides a method of determining the number of charge carriers per unit volume. With semiconductors, the direction of V_{H} indicates whether the majority carriers are positive holes or negative electrons. A calibrated *Hall probe* gives an immediate read-out of magnetic field strength.

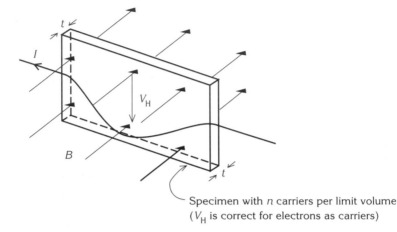

Specimen with n carriers per limit volume
(V_{H} is correct for electrons as carriers)

Hall probe: an instrument, based on the *Hall effect*, for measuring *magnetic field strength*. The probe is attached to a calibrated meter to give an immediate read-out. To determine the magnetic field strength at a point, the probe should be placed at the point and twisted and turned until the maximum read-out is obtained. The meter reading gives the magnitude of the magnetic field strength and its direction will be at right angles to the plane of the probe. See *Hall effect*.

hardness is a measure of how difficult it is to scratch a material. Scales of hardness are built up either by testing which materials scratch which (squirrels' teeth are harder than steel) or by pressing the surface with a standard force on a hard vertical point and measuring the depth of any depression.

heat is energy moving down a temperature gradient. (See *heat capacity, thermal conduction, heat transfer.*)

heat capacity is the *heat*, in joules, needed to raise the temperature of a body by one degree kelvin. The SI unit is joules per kelvin, J K^{-1}. The symbol used is C (note the upper case form). The word equation definition is:

$$\text{heat capacity} = \frac{\text{heat transferred}}{\text{temperature rise}}$$

$$\text{J K}^{-1} = \frac{\text{J}}{\text{K}}$$

$$C = \frac{\Delta Q}{\Delta \theta}$$

The heat transferred must cause a temperature rise and nothing else. There must be no change of state.

Heat capacity is a physical property of a body not of a material. A bus has a heat capacity. If the body is a homogeneous lump of one material, the heat capacity divided by the mass (the heat capacity per unit mass) is called the *specific heat capacity* of the material. Specific heat capacity is a property of a material, not of a body.

Worked example

A tea urn is equipped with a 3.0 kW heating element. When it is half-filled with water the temperature rises 55 K in 11 minutes. Calculate the heat capacity of the tea urn.

$$C = \frac{\Delta Q}{\Delta \theta} = \frac{(3.0 \times 10^3 \text{ W})(11 \times 60 \text{ s})}{(55 \text{ K})} = 36 \text{ kJ K}^{-1} \quad \text{(to 2 s.f.)}$$

How far this heat capacity belongs to the urn and how much of it belongs to the water cannot be decided without more information.

heat engine: a mechanism which receives *heat* from a hot 'source' and passes some of this heat to a cooler 'sink' while converting the remainder into useful *work*. (See also *heat engine concept*.)

heat engine concept: the abstract model of a *heat engine* illustrated in the diagram below. The engine takes in heat Q_1 from the hot 'source' whose temperature is T_1. It passes heat Q_2 to a second body, called the 'sink', at temperature T_2. T_1 and T_2 are temperatures on the absolute scale. The engine does work W on its surrounding, where $W = Q_1 - Q_2$.

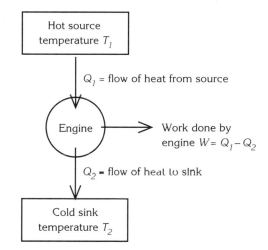

The efficiency of a heat engine is defined as:

$$\text{efficiency} = \frac{\text{work done by engine}}{\text{heat taken in from source}}$$

$$\eta = \frac{W}{Q_1}$$

where η is the efficiency of the engine. This equation is a definition; it applies to any engine no matter how inefficient it may be.

A heat engine which can be run backwards, that is, can be worked upon to absorb heat Q_2 from the cold sink and reject heat Q_1 into the hot source, is called a reversible heat engine. The *second law of thermodynamics*, applied to a reversible heat engine, shows that its efficiency sets an upper limit for all heat engines. This efficiency comes out as follows:

$$\eta \;=\; \frac{Q_1 - Q_2}{Q_1} \;=\; \frac{T_1 - T_2}{T_1} \;=\; \frac{W}{Q_1}$$

This equation shows that high efficiency in real heat engines is obtained by ensuring a large temperature difference between source and sink. See *engine cycles, heat pump, refrigerator.*

heat pump: a *heat engine* which is worked backwards and which thereby removes internal energy Q_2 from a cold region and discards heat Q_1 to a hotter region where Q_1 equals the sum of Q_2 and the thermal equivalent of the work done on the heat engine, i.e. $Q_1 = Q_2 + W$.

Heat pumps are used for domestic heating and the emphasis is on the heat Q_1 discarded into the house, not on the heat Q_2 removed from the cold exterior. For this reason, the coefficient of performance for a heat pump is defined differently from the same term used with refrigerators:

$$\text{coefficient of performance} \;=\; \frac{\text{heat discarded at high temperature}}{\text{work done by engine}}$$

$$=\; \frac{Q_1}{W} \;=\; \frac{Q_1}{Q_1 - Q_2} \left\{ = \; \frac{T_1}{T_1 - T_2} \right\}$$

If the engine is less than perfect, W is proportionately too large and the part of the expression involving absolute temperatures (the part in the large brackets) no longer holds. A refrigerator is a heat pump in which the emphasis is on the heat Q_2 removed from the low temperature container. (See also *heat engine concept, refrigerator.*)

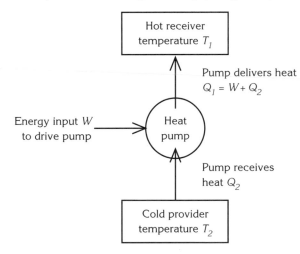

heat transfer is the flow of energy away from a body into the surroundings by thermal conduction, convection, heat radiation or evaporation.

heating is the flow of disordered molecular kinetic energy into a body, thereby increasing its internal energy and raising its temperature. The two conditions which are necessary for heating to occur are a temperature difference and thermal contact. The warmer of the two bodies heats the cooler body. Thermal contact includes the idea of a pathway for radiation to travel from a hot source like the sun to the cooler body of interest.

heating and working are two forms of energy transfer; heating is energy transfer driven by a difference in temperature whilst working is energy transfer driven by a force of some description.

helical spring: a spring formed by winding a wire round and round a uniform cylinder. The usefulness of a helical spring is linked to its main property: its extension is proportional to the load applied. This relationship is illustrated by the graph in the diagram.

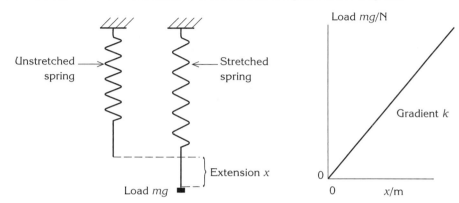

The proportionality of extension to load is called *Hooke's law*. It can be stated in algebraic form:

$$F = kx$$

where k is called the spring constant, F is the stretching force and x is the extension. A typical spring in a school laboratory will be stretched by about 5 cm by the weight of a 100 g mass. The weight of a 100 g mass is about one newton and 5 cm is a twentieth of a metre, so the resulting spring constant is in the region of 20 N m^{-1}. It can be checked by suspending the spring from a rigid support and measuring carefully the extensions corresponding to different loads. The gradient of the load/extension graph is the spring constant k.

The energy stored in the extended spring equals the work done stretching it and this equals the area between the load/extension graph and the extension axis. (Don't write 'the area under the curve'. If you have the axes the wrong way round and if the graph is not straight, the two areas between the graph line and the axes will be different.) The area between the curve and the extension axis is the average force multiplied by the extension:

energy stored in spring = work done in stretching the spring

= average force × extension

= (½ maximum force) × extension

= (½ × kx) × x

= ½ kx^2

If the mass supported by the spring is raised slightly and then released, the mass oscillates up and down with *simple harmonic motion*. It is simple harmonic because the resultant force on the mass is the difference between its weight and the tension in the spring and this resultant force is proportional to the displacement, s say, of the mass from its rest position. We could write:

$$F = -ks = ma$$

or

$$a = -\frac{k}{m}s = -\frac{g}{x}s$$

since $mg = kx$. The negative sign has appeared because s is a small displacement upwards but it is measured in a downwards direction. This is the direction in which F, x and a are measured.

The two equations for acceleration a describe simple harmonic motion of period T, where:

$$T = 2\pi\sqrt{\left(\frac{m}{k}\right)} = 2\pi\sqrt{\left(\frac{x}{g}\right)}$$

When a helical spring is stretched by a mass in a gravitational field, and when the mass is oscillating up and down, energy is stored in three ways:

- the *kinetic energy* E_k of the oscillating mass
- the *gravitational potential* energy E_p of the mass
- the energy W stored in the spring.

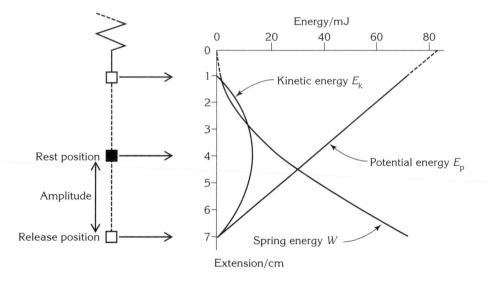

The relationship between these three quantities of energy is shown in the second graph. The values on the graph apply to a 0.120 kg mass hung from the lower end of a helical spring. The extension is 0.040 m. The mass is then pulled down 0.030 m and released.

henry (H): the SI unit for *self-inductance* and *mutual inductance*.

Hertzsprung–Russell diagram: a scatter diagram of absolute magnitude against spectral type for a large number of stars. Alternatives to spectral type are temperature T and colour index $M_B - M_V$. If spectral type is spread out evenly, then the temperature scale will be non-linear. The luminosity ratio L/L_{Sun} , expressed as a power of ten, can be used instead of absolute magnitude. These different possibilities are shown on the diagram below.

The shaded line from top left to bottom right represents stable stars; it is called the main sequence. Theory shows that luminosity for these stars increases with stellar mass and the relationship is verified by those stars whose masses are known. The relative mass scale on the left applies to main sequence stars only. To the left of this scale are marked the possible ends states. The scale on the extreme left shows the relative numbers of stars of different masses. Notice how small is the proportion of very large stars.

One useful feature of the diagram is that the luminosity L of a distant star can be read from the diagram once the temperature of the star is known from Wien's law. The intensity I of the same star can be measured using a telescope with a radiation sensor. The distance D of the star can now be calculated from the relation

$$I = L/4\pi D^2.$$

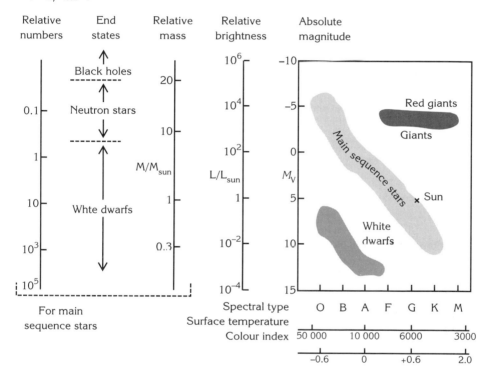

| Relative numbers | End states | Relative mass | Relative brightness | Absolute magnitude |

For main sequence stars

Spectral type O B A F G K M

Surface temperature

Colour index

high-voltage transmission is a technique for reducing the power loss incurred when electrical power is transferred across large distance through power transmission lines. The trick is to have a step-up transformer at the input end of the transmission line and a step-down transformer at the output end. Imagine that electrical power is generated at 25 kV and supplied to the consumer at 240 V. The input transformer can step up the voltage from 25 kV to 400 kV. Employing the transformer equation:

$$I_p V_P = I_s V_S$$

we see that the current in the transmission lines will be one sixteenth of the current in the generator output circuit. But power loss in a cable is I^2R where I is the r.m.s. current in the cable and R is the resistance of the cable. Reducing the transmission current to one sixteenth of its former value, reduces the power loss in the cable by a factor of 256. The gain is not in practice as dramatic as this, but it is substantial. The voltage at the consumer end will be cut down in stages to ensure similar savings in transmission at the local level.

homogeneity is a necessary condition for an equation involving physical variables to be valid. It is not the only condition; the numbers must also balance. The equation:

four horses = seven horses

is wrong because the numbers don't balance. It is homogeneous because it has 'horses' on both sides. The equation

four horses = four castles

is wrong because horses and castles are altogether different. The numbers balance but it is not homogeneous. Homogeneity means that the two sides of an equation correspond to the same kind of physical quantity. The equation:

four horses = four horses

is correct since the numbers balance and it is homogeneous.

It may be that an equation has two or more terms on one side, as in:

$$3x + 2y = 5z$$

For this equation to be homogeneous, x, y and z must all be the same kind of thing, e.g. lengths.

Hooke's law states that stress is proportional to strain. It applies to elastic substances within the limit of proportionality. (See *helical spring*, *stress–strain curve*.)

Hubble's law states that the speed of recession, v, of any one galaxy from another is proportional to their distance apart, d. The constant of proportionality H ($= v/d$), known as the Hubble constant, is between 50 and 100 km s^{-1}/Mpc. H is found by measuring the speed of recession for those galaxies whose distances are known. The speed of recession is measured using the *Doppler effect*. Galactic distance is measured in a number of ways. If, for example, a *cepheid variable* star is present in the galaxy, its peak brightness can be determined from its period and its apparent brightness can be measured directly. These two quantities give the distance.

Whether or not the remote galaxies will eventually slow down, reverse direction and trigger another big bang some long time into the future depends on their speeds and the average density of the universe. At the moment, this average density is not known with sufficient accuracy to foretell the future.

hydrogen atom: an atom constructed from a single *electron* and a single *proton*. The hydrogen atom is an important element in physical theory because it has these two components and no more. This simple structure is amenable to extensive and detailed analysis. Quantum theory can account for the physical properties of the isolated hydrogen atom and for chemical properties of hydrogen in combination with other atoms. This successful analysis is an essential springboard to the study of more complex atoms.

Following Rutherford's discovery of the atomic nucleus, Bohr developed a simple model of the hydrogen atom which embodied the all-important quantum principle and accounted for the wavelengths of the lines of the hydrogen spectrum with impressive accuracy. See *nuclear atom*.

hysteresis is a term used with elastic and magnetic materials whereby the path followed to elastic deformation or magnetisation differs from the path for stress release or for demagnetisation. In both cases it implies that rapid repetition of the stress (or the magnetisation) process results in a heating of the specimen. The elasticity effect is shown especially by rubber and other polymers. In elastic hysteresis, the energy linked to the strain corresponding to the stress, is released before the strain is wholly removed. In magnetic hysteresis the effect works the other way, the material keeps some of its magnetisation when the magnetising field is switched off.

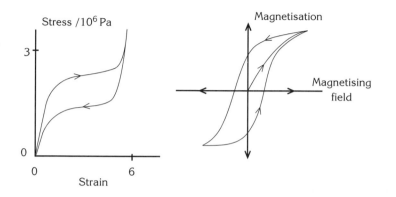

(See *stress–strain curves*.)

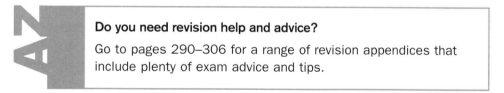

Do you need revision help and advice?

Go to pages 290–306 for a range of revision appendices that include plenty of exam advice and tips.

ice point: the temperature of a pure ice and water mixture in a thermal equilibrium state at standard atmospheric pressure. It is 273.15 K.

The definition of the *triple point* leaves out the reference to standard atmospheric pressure and adds water vapour to the equilibrium mixture.

ideal gas equation: an equation which describes an equilibrium state of an ideal gas and which is usually presented in the form:

$$pV = \frac{m}{M}RT = nRT$$

where:

p is gas pressure in pascals (Pa)
V is the volume of gas in cubic metres (m³)
m is the mass of gas in kilograms (kg)
M is the molar mass in kilograms per mole (kg mol⁻¹)
n is the amount of gas in moles (mol)
R is the molar gas constant whose magnitude is 8.31 J K⁻¹ mol⁻¹
T is the gas temperature in kelvins (K).

The ideal gas equation is suggested by experiments with real gases and many real gases and mixtures, like air and helium obey it to a high degree of accuracy. But an ideal gas remains a concept, not a reality. It is a standard to which the behaviour of real gases can be approximated.

Of the five or six physical variables which appear in the equation, all but one (temperature T on the *absolute thermodynamic temperature scale*) are experimental quantities which can be measured for real gases. The constant volume hydrogen thermometer gives values which can be corrected to the absolute scale and be given in kelvins.

When the equation is applied to real gases, they must be in an equilibrium state; all the variables must have settled down to their equilibrium values. The equation applies to a gas before and after a change but not during the change. Also, like any other equation we use, it is valid only if we remember to use SI units.

Worked examples

1. The molar mass of air is 0.029 kg mol⁻¹. Find the density of air at atmospheric pressure (101 kPa) and room temperature (25°C).
    We must use temperatures on the absolute scale,
    $T = 273$ K $+ 25$ K $= 298$ K.

Density is mass/ volume, m/V.

$$\text{density} = \frac{m}{V} = \frac{pM}{RT}$$

$$= \frac{(1.01 \times 10^5 \text{ Pa})(0.029 \text{ kg mol}^{-1})}{(8.31 \text{ J K}^{-1} \text{ mol}^{-1})(298 \text{ K})}$$

$$= 1.2 \text{ kg m}^{-3} \quad \text{(to 2 s.f.)}$$

2. A 250 cm³ volume of air at 100 kPa pressure and exactly 0°C is allowed to settle down in a space of 320 cm³ at 30°C; find the new pressure.

    Since the question refers to a constant mass of gas, we can use a reduced form of ideal gas equation:

$$\frac{P_1 V_1}{T_1} = \frac{P_2 V_2}{T_2}$$

There are four points which we must remember about this equation:

- the states, before and after, must be equilibrium states
- the mass of gas must be the same for both states
- pressures (or volumes) do not have to be in SI units but must be in the same units
- temperature must always be in kelvins.

We have then:

$$P_2 = P_1 \frac{V_1 T_2}{V_2 T_1} = (100 \text{ kPa})\frac{(250 \text{ cm}^3)(303 \text{ K})}{(320 \text{ cm}^3)(273 \text{ K})} = 87 \text{ kPa} \quad \text{(to 2 s.f.)}$$

(See also *Boyle's Law, Charles' Law, absolute temperature scale*.)

impulse is the product of an average force and a time interval, it is not a kind of force. When a tennis ball is struck with a racket, the force on the ball changes magnitude rapidly and the impact is all over in a moment. How can we deal with such a situation? At any instant during the collision the force the racket applies to the ball is matched by the reaction force from the ball on the racket. Represent these two forces with symbols $F_{r \to b}$ for the force of the racket on the ball and $F_{b \to r}$ for the reaction force of the ball on the racket.

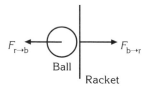

We have, from *Newton's third law of motion*, the relationship:

$$F_{r \to b} = -F_{b \to r}$$

If these force are active for time dt seconds, then:

$$F_{r \to b} dt = -F_{b \to r} dt$$

But, we know from *Newton's second law of motion*, that:

$$\text{resultant force} \times dt = d(mv)$$

It follows that:

$$F_{r \to b} dt = \text{momentum transferred to the tennis ball in time } dt$$
$$= - \text{momentum transferred to the tennis racket in time } dt$$

If all the small changes in momentum are added up for the whole period of contact, then, using the symbol Σ for 'the sum of', we have:

$$\Sigma F_{r \to b} dt = \text{change in momentum of tennis ball during the strike}$$
$$= (\text{average value of force of racket on ball}) \times \text{contact time}$$
$$= \text{impulse of racket on ball}$$

(The second line is, in fact, the definition of average force.)

The word equation definition for impulse is:

impulse = momentum transferred during contact in a collision

and it equals average force $\times$ contact time.

Given this the SI unit for impulse must be the same as the SI unit for momentum, i.e. newton second (N s).

If a graph is drawn of force against time, then impulse equals the area between the curve and the time axis.

A working definition of impulse is the product of average force and contact time. Therefore, for a given impulse, increasing the contact time lowers the average force.

Worked example

A cricket ball of mass 145 g travelling at speed 35 m s^{-1} is caught by a fielder and brought to rest in 0.080 s. Calculate the average force applied to the ball and the distance over which this force acts.

$$\text{average force} = (\text{momentum change}) \div \text{time}$$
$$= (0.145 \text{ kg} \times 35 \text{ m s}^{-1}) \div (0.080 \text{ s}) = 63 \text{ N} \quad (\text{to 2 s.f.})$$
$$\text{distance} = (\text{loss of kinetic energy}) \div (\text{average force})$$
$$= (\tfrac{1}{2} \times 0.145 \text{ kg} \times (35 \text{ m s}^{-1})^2) \div 63 \text{ N} = 1.4 \text{ m} \quad (\text{to 2 s.f.})$$

induced e.m.f.: an *e.m.f.* generated in a conductor either by a changing magnetic flux around the conductor or by the conductor itself cutting across the magnetic field. (See also *back e.m.f., electromagnetic induction, laws of electromagnetic induction.*)

inductance: see *self-inductance, mutual inductance.*

inductive reactance is the ratio of peak voltage to peak current for an inductor. Some care is needed; these two quantities peak at different times.

The diagram on page 135 shows an inductor of inductance L connected across a power supply of output voltage V equal to $V_0 \sin 2\pi ft$ or $V_0 \sin \omega t$. The angular frequency ω radians per second equals $2\pi f$ where f is the frequency in cycles per second. It is assumed that the resistance of the coil is zero.

The behaviour of the inductor is governed by the equation:

$$E = -L\frac{dI}{dt}$$

Where E is the e.m.f. induced in the inductor when the current is changing at the rate dI/dt. But

$$V = V_0 \sin \omega t$$

and, since E is the only voltage available to oppose V,

$$V = -E = L\frac{dI}{dt}$$

If we put

$$\frac{dI}{dt} = \frac{V}{L} = \frac{1}{L} V_0 \sin \omega t$$

then

$$I = -\frac{1}{\omega L} V_0 \cos \omega t$$

$$= -I_0 \cos \omega t$$

where

$$I_0 = \frac{1}{\omega L} V_0$$

The negative sign in the expression for the current I means that the current I peaks after the voltage V peaks and not before. It is all a consequence of V balancing E which is in turn proportional to the rate of change of current and not proportional to the current itself.

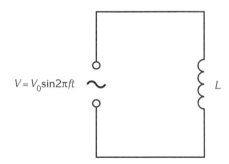

$$V = V_0 \sin 2\pi ft \qquad L$$

Compare this with Ohm's law:

$$V = IR \qquad\qquad V_0 = \omega L I_0$$

In Ohm's law, the resistance R holds the balance between current and voltage. For the inductor, ωL holds the balance between peak voltage and peak current but the quantities are not in phase. This is why the negative sign appears in the equation relating current to voltage. The expression ωL is called the reactance of the inductor. The symbol for inductive reactance is X_L and its SI unit is the ohm (Ω).

The best way of seeing what all this means is by working carefully through an example.

135

Worked example

Let the power supply have a peak output V_0 of 5.0 V, let its frequency f be 1.00 kHz (ω equals $2\pi f$ or 6.28×10^3 rad s^{-1}) and let L equal 100 mH. Assume that the inductor has no ohmic resistance. Start by finding the reactance of the inductor:

$$X_L = \omega L = (6.28 \times 10^3 \text{ rad s}^{-1})(0.100 \text{ H}) = 628 \text{ }\Omega$$

With peak voltage at 5.0 V, the peak current is 5/628 or 8.0 mA (to 2 s.f.). Equations for voltage and current can now be written out in detail and plotted on a graph:

$$V = 5.0 \sin (6.28 \times 10^3 \times t)$$
$$I = -(8.0 \times 10^{-3}) \cos (6.28 \times 10^3 \times t)$$

(6.28×10^3) is a large number and this is why the time axis on the following graph is in milliseconds. Notice that the current axis is calibrated on the left-hand side of the graph and the voltage is calibrated on the right-hand side.

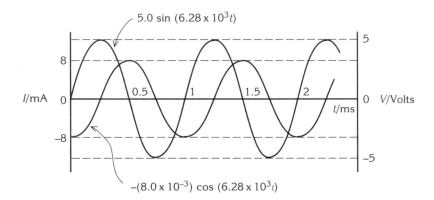

The time axis is drawn with time increasing towards the right. The current in the inductor peaks in the same direction as the voltage a quarter of a cycle later than the voltage. This explains the phrase 'current lags voltage'. The current would need a few cycles to settle down at the values shown in the graph.

(See also *reactance, series resistance and capacitance, capacitive reactance, series resistance and inductance, resonance in a.c. circuits*.)

inelastic collision: a collision in which *kinetic energy* is not conserved. *Linear momentum* will be conserved.

infrared radiation: see *IR*.

insulators are materials which contain no charge carriers free to move within their volumes. Electrical insulators cannot carry electrical current. Even so, the insulating properties of many materials break down if too high a voltage is applied. Note the contrast with good thermal conductors and poor thermal conductors. (See *conductors, resistivity*.)

intensity (stellar): the radiant power per square metre reaching the Earth's surface from a star. If D is the distance of the star from the Earth and if L is its *luminosity*, then $I = L/4\pi D^2$. See *star*.

interference is the formation of a new wave pattern where two or more incident wave patterns overlap. It is a visible phenomenon. Young's fringes can be seen by someone who has no knowledge of physics. The interpretation of an interference pattern, the explanation of how it happens, is convincing evidence of the wave character of the incident beams.

Interference is a consequence of the *principle of superposition* and happens provided that certain conditions are met. (See also *constructive interference, destructive interference, coherence, two slits experiment.*)

internal energy is the sum total of the kinetic and potential energies of all the individual molecules which go to make up an object. In an ideal gas there are no intermolecular forces and no intermolecular potential energy so the internal energy is kinetic only. Internal energy decides the temperature and state of a body; an increase is accompanied either by an increase in temperature or by a change of state, (e.g. melting). In practice, changes of internal energy are more significant than attempts to evaluate the actual level of internal energy. Internal energy can be increased in either or both of two ways; by working on a body or by heating it. (See *heat capacity, enthalpy, heating and working, thermodynamics (first law of).*)

internal resistance is the resistance to current flow in an electrical power source. It is a topic which needs to be approached with a little care. Resistance is defined as the ratio of the potential difference V across a conductor to the current I in the conductor, $R = V/I$. It is not possible to use this equation to define internal resistance because the voltage drop V_r across the imagined internal resistance r is not accessible. The best we can do is to write down the equation:

$$E = I(R + r)$$

where E is the e.m.f. of the cell, R is the external resistance, r is the internal resistance and I is the current in the circuit. We then record a series of measurements of the current I for different values of the external resistance R and draw a graph of $1/I$ against R:

$$\frac{1}{I} = \frac{R}{E} + \frac{r}{E}$$

The graph is a straight line of gradient $1/E$ and intercept r/E on the $1/I$ axis.

A clearer idea of the meaning of internal resistance is arrived at by multiplying the equation for internal resistance through by the current I, leading to the equation:

$$E I = I^2 r + I^2 R$$

EI is the energy transformed into electrical energy in the power source, I^2R is the work done by the power source on the external circuit and I^2r is the work done within the power source.

When internal resistance is measured, there appears to be a choice between taking a series of measurements of current and resistance or a series of measurement of current and voltage. In fact, standard resistors are expensive devices which are meant to be operated at low current. But internal resistances are often small, often a fraction of an ohm. In such cases an appreciable current must be drawn from the cell or power source to enable the internal resistance to be calculated with reasonable accuracy. For this reason, it is better to use a rugged, reliable and variable external resistance (whose precise magnitude we do not need to know) in conjunction with a digital voltmeter and a digital ammeter.

Experiment: to measure the internal resistance of an electrical power source

Apparatus

A power source, e.m.f. E and internal resistance r, is connected to an external variable resistor of resistance R. The current I in the external resistance and the potential difference V across it are measured with a digital ammeter and a digital voltmeter. The circuit is shown in the diagram.

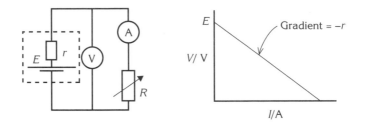

Measurements

The potential difference V is read from the digital voltmeter and the current I is read from the digital ammeter for each of six values of the variable resistor. The resistance values should be arranged such that the voltage V should drop from a maximum value almost equal to E to a minimum value about a third lower. A larger drop in voltage would mean drawing a larger current than would be good for the power source.

Conclusion

The voltage V equals the e.m.f. E of the power supply less the voltage drop $I\,r$ across the internal resistance:

$$V = E - I\,r$$

A graph of the voltage V against the current I will be a straight line of gradient $-r$ starting from the point E on the voltage axis. The measurements will cover about the top third of this line; the rest of the line is an extrapolation.

Comment

The restricted set of voltage values look like a fault in the design of the experiment. In fact they ensure that measurements are taken within the range of current values that the power supply would be called upon to deliver. Also, the internal resistance would probably increase if too large a current was drawn from the supply.

International Practical Temperature scale: a temperature scale which is used for calibrating thermometers and which is as close as possible, numerically, to the absolute thermodynamic temperature scale (see *absolute temperature scale*). It is defined in two stages:

- a large number of fixed points are mapped onto particular numbers on the scale. The *triple point*, for instance is mapped onto 273.16 K.

- the specifications for constructing thermometers to measure temperatures between the fixed points are defined (for instance, the design of a thermocouple to measure temperatures between the ice point and the steam point).

ionic bond is a form of attachment between the atoms in a molecule in which electrons from an atom of one element are transferred to the atom of a second element. The two atoms are now oppositely charged. An ordered structure can be built up in which the oppositely-charged atoms of the two elements alternate in all three directions to give a crystalline solid.

The ionic bond is not as strong as the *covalent bond* and, since it relies on the electrical attraction between neighbouring atoms, it breaks down in the presence of polar solvents such as water. The weakness of the bond gives us a solid which is not very hard and which either has a low-to-modest melting point or dissociates on heating. The neutralising of the inter-atomic attraction in the presence of water implies a strongly soluble substance. See *metallic bond*.

ionisation is the process by which an atom receives enough energy to lose at least one electron. (See *energy level*.)

ionising radiation is a term which covers any kind of radiation which will ionise a gas put in its path. This includes *alpha*, *beta* and *gamma radiation* in addition to *X-rays*.

IR (infrared radiation) is that segment of the electromagnetic spectrum which lies between the visible and the microwave regions. It is primarily associated with hot surfaces. Human skin senses it easily but there is little resolution.

IR wavelengths are from 700 nm (0.7 µm) up to about 1 mm. This is so broad that it is usual to divide it into the near IR (0.7 µm to about 3 µm), thermal IR (3 µm to 10 µm) and the far IR to cover the remainder. This division is essential to the study of the Earth by remote sensing techniques since they relate to different kinds of source and treat different areas of information. Near IR can be detected with a photocell or photo-diode but the rest of the spectrum is investigated with a bolometer (a thin strip of blackened platinum creating a kind of resistance thermometer), a *thermocouple* or a thermopile (lots of thermocouples in series).

isothermal changes are changes at constant temperature. A fixed mass of gas obeys *Boyle's law* when undergoing an isothermal change. It is sometimes convenient to resolve changes in the state of a gas into two changes; one *adiabatic*, the other isothermal. This principle provides the theoretical basis for the tephigrams used in meteorology.

isotope: a nuclide with the same *atomic number* as another nuclide but with a different *mass number*.

jet engines create a forward thrust on an aeroplane by the explosive discharge of hot burnt fuel through the rear of the engine. The momentum of the burnt fuel discharged per second towards the rear equals the increased momentum of the aeroplane in the forward direction. The design of the engine is complicated by the need to access the surrounding air for the oxygen in which the fuel burns. As a consequence, jet engines cannot fly very high. Rockets carry their own supply of oxygen or whatever and can operate in space.

joule: another name for the newton metre. It is the SI unit for energy and its symbol is J. The *base unit* equivalent of the joule is $kg\ m^2\ s^{-2}$. (See also *work, derived unit, kilowatt-hour*.)

Joule–Thomson effect: the name given to the slight cooling of a gas when it expands slowly without working on its surroundings. The original experiment involved pushing a gas through a cotton wool plug against a second piston on the far side of the cotton wool. The pressures on the two pistons were kept steady at values which ensured that the work done by the gas against the second piston equalled the work done on the gas by the first piston. The net work done by the gas was zero but its density fell slightly, as it moved through the cotton wool and its molecules moved slightly apart. The cooling effect was a measure of the weak but still-significant attractive forces between the molecules of the gas.

If the gas were perfect there would be no residual molecular attractions and no cooling effect. As it was, the observed cooling effect represented a departure from perfect gas behaviour. This departure enabled Thomson (later to become Lord Kelvin) to correct the readings of a constant volume hydrogen thermometer to give temperatures that a constant volume thermometer filled with ideal gas would have given had it been available. This is what made the absolute scale of temperature a practical possibility. Joule did all the experimental work.

The discovery of the cooling effect had other unforeseen outcomes. Air which is cooled down by pumping it through a small hole can be used to cool down the incoming gas. Eventually the air will cool down to the point where it liquefies.

K-capture means the capture by the nucleus of an atom of an electron from the atom's K-shell. The K-shell is the shell nearest to the nucleus and contains the most tightly bound electrons. K-capture is an alternative to β^+ decay and much more common. An example is the decay of beryllium-7 into lithium-7:

$$_4^7Be + {}_{-1}^0e \longrightarrow {}_3^7Li + {}_0^0v$$

The neutrino and the electron have the same lepton numbers; lepton number is conserved. The captured electron combines with a proton in the nucleus to form a neutron:

$$_1^1p^+ + {}_1^0e \longrightarrow {}_0^1n + {}_0^0v$$

kelvin: the SI unit of temperature difference. The symbol is K. It is defined as the temperature difference between *absolute zero* and the *triple point* of water divided by 273.16. Temperatures measured from absolute zero are expressed in kelvins (K). Temperatures measured from the freezing point of water (273.15 K) are expressed in degrees Celsius (°C). Temperature differences measured on either scale are given in kelvins (K).

It was Lord Kelvin who showed that the theory of heat engines led to the definition of an absolute thermodynamic scale of temperature which was independent of the properties of any real substance. He showed further that this scale of temperature was identical to the ideal gas scale of temperature. He went on to show how temperatures measured with a constant volume gas thermometer filled with hydrogen could be corrected to temperatures measured on the absolute scale.

Kepler's laws are three laws of planetary motion which were discovered empirically by Kepler early in the seventeenth century and shown later in the same century by Sir Isaac Newton to follow from the law of universal gravitation. They are as follows:

1. A planet moves round the sun along an elliptical path with the sun at one focus of the ellipse.

2. The straight line joining a planet to the sun sweeps out equal areas in equal times.

3. The ratio of the square of a planet's orbital period to the cube of the major diameter of its orbital path is the same for all planets.

kilowatt–hour: the energy transferred from a source which delivers energy for one hour at a constant rate of a thousand watts. It is not an accepted SI unit, but is widely used as a practical unit (by energy suppliers for example), has the symbol KWh and equals exactly 3.6 MJ.

kinetic energy is the energy possessed by a mass, by reason of its motion. It is a *scalar quantity*. The SI unit is the *joule* (J). The symbol used may be the energy symbol *W* or there may be a subscript (W_k) to distinguish it from potential energy W_p. The symbol E_k is also used. (Some writers use k.e. and even K.E.)

For mass m travelling at speed v:

$$\text{kinetic energy} = \tfrac{1}{2}mv^2$$

You need to be familiar with 'work done equals increase in kinetic energy' calculations, with the interplay between *potential energy* and kinetic energy in gravitational fields and with *simple harmonic motion* (this is straightforward with the *simple pendulum* but more complicated with the *helical spring* where potential energy is stored in two forms).

Thermal energy is wholly kinetic in gases. It is partly kinetic and partly potential in liquids and solids where inter-atomic distances are small.

kinetic theory of gases: an explanation of why gases obey the gas laws, it is based on the idea of a gas as a set of hard particles (this is called the model) which obey *Newton's laws of motion*. The theory is developed in stages:

- write out a set of postulates which define the model
- apply Newton's laws of motion to this model and so relate pressure to average kinetic energy
- assume that average *kinetic energy* is proportional to *absolute temperature*
- confirm that the theory leads to a full statement of the *ideal gas equation*.

Postulates:

- a gas consists of a large number of identical particles called molecules
- molecular diameter is much smaller than the average distances between the molecules
- molecules obey Newton's laws of motion
- molecular speeds and directions are distributed randomly
- there are no forces between molecules except when they collide
- collisions are elastic and instantaneous.

Applying Newton's laws of motion:

Imagine N molecules of mass m' occupying a cube of side l with hard walls:

the momentum change at the wall resulting from the reflection of a molecule with speed v_1 normal to the wall = $2m'v_1$ (see the diagram on page 143).

$$\text{momentum change per second} = 2m'v_1 \times \text{number of collisions per second}$$
$$= 2m'v_1 \times [v_1/2l] = m'v_1^2/l$$
$$= \text{force on one wall from one molecule}$$

Since, for each molecule,

$$u_1^2 + v_1^2 + w_1^2 = c_1^2,$$
$$\text{the total force on all six walls} = 2 \times [m'/l] \times [c_1^2 + c_2^2 + c_3^2 \ldots c_N^2]$$
$$= [2m'/l] \times [N{<}c^2{>}]$$
$$\text{where} \quad {<}c^2{>} = [c_1^2 + c_2^2 + c_3^2 \ldots c_N^2]/N$$

${<}c^2{>}$ is the average value of (molecular speed^2). It is called the *mean square speed*.

We now have:

pressure p on wall = total force / area

$$= [2m'/\, l\,] \times [N\!<\!c^2\!>] \div [6\,l^2] = [Nm'/3V]\,<\!c^2\!> = p$$

and since Nm'/V is the gas density ρ, we have:

$$p = \tfrac{1}{3}\,\rho\!<\!c^2\!>$$

or,

$$pV = \tfrac{1}{3}\,Nm'\!<\!c^2\!>$$

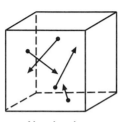

N molecules;
different speeds
different directions

$\Delta(mv) = 2mv$

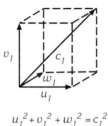

$$u_1^2 + v_1^2 + w_1^2 = c_1^2$$

Assumption concerning temperature:

The next step is to involve temperature by assuming that average molecular kinetic energy is related to temperature in the following manner:

$$\tfrac{1}{2}m'\!<\!c^2\!> = \tfrac{3}{2}\,kT$$

(The strict statement of the assumption is that each of the three components of average molecular kinetic energy is $\tfrac{1}{2}\,kT$)

General gas equation:

If we put this assumption in our expression for gas pressure, we have:

$$pV = (\tfrac{2}{3}N)(\tfrac{1}{2}\,m'\!<\!c^2\!>) = NkT$$

where k is the Boltzmann constant, 1.38×10^{23} J K^{-1}.

Next, note that

$$N = \frac{m}{M}\,L = nL$$

where m is the mass of the gas in kilograms, M is the molar mass, L is the Avogadro constant (6.022×10^{23} mol^{-1}) and n is the number of moles. And if we put:

$$R = kL = (1.38 \times 10^{-23} \text{ J K}^{-1})(6.022 \times 10^{23} \text{ mol}^{-1}) = 8.31 \text{ J K}^{-1} \text{ mol}^{-1}$$

then our equation becomes:

$$pV = \frac{m}{M}\,RT = nRT$$

where n is the number of moles of gas in the sample.

Kirchhoff's first law

The derivation of this law confirms our confidence in the molecular theory of matter. The next stage would be either to introduce other postulates to account for deviations from the gas laws at high densities and pressures, or to explain other phenomena such as diffusion rates or thermal conductivity. But none of this is in the syllabuses. (See also *ideal gas equation*.)

Kirchhoff's first law states that the total electric current flowing into a junction along all possible roots is zero. An alternative form states that the sum of the currents flowing towards the junction of a number of conductors equals the sum of the currents leaving the junction. Either way, the implication is that electric charge is a conserved quantity.

The law may seem an empty and even pointless statement to someone in the early stages of studying electricity but it is an essential step in setting up a formal algebra for circuit analysis. There is a hint of this formalism in the derivations of the rules for combining *resistors in series* and for combining *resistors in parallel*.

Kirchhoff's second law states that the net *e.m.f.* in a circuit loop equals the sum of the *potential differences* round the loop. E.M.F.s are sources of electrical energy, potential differences are drains of electrical energy. Kirchhoff's second law is a statement of the law of *conservation of energy* in a form which applies to electrical circuits. The net e.m.f. E is the electrical energy provided by the power source per coulomb of charge flowing in the circuit. Each potential difference states how much electrical energy is dissipated, per coulomb of charge flowing, in the element to which the potential difference applies. The law tells us that, for any complete loop in a circuit:

$$E = V_1 + V_2 + V_3 \ldots V_n$$

Do you need help with the synoptic element of your Physics course?

Go to page 296 for tips and advice.

laminar flow is an alternative name for *streamline flow*.

length is a *base quantity* in SI. It is a property of a line. The length of a line, whether or not it is straight, is measured with a ruler or its equivalent. The SI unit of length is the metre (m). (See also *distance, distance travelled, displacement*.)

length, distance and displacement are three closely-related concepts. Length is part of our language. If length has a definition, it is the outcome of measuring a length with a ruler. Length is the physical quantity: it is the magnitude that is being measured. It may be the length of a conducting wire, the length of a solenoid or the length of a river. The unit is the metre.

Distance is the length of a pathway between two points. Imagine that a string is laid along the pathway, marked at each of the two points and then pulled out straight. The distance between the two points on the pathway is the length of the straightened string between the two marks. Distance and length are *scalar* quantities.

Displacement is a *vector* quantity which defines the position of a second point relative to a fixed point. No pathway is involved. The magnitude of the displacement is the straight line distance between the two points and its direction is the angle between this straight line and a reference line. These two lines must start at the first point.

[The position vector in the vector work for mathematics A level is the displacement of a point from the origin of the co-ordinate system in use.]

lens equation is the equation:

$$\frac{1}{u} + \frac{1}{v} = \frac{1}{f}$$

where u is the distance of the object (the light source) from the lens, v is the distance of the image from the lens and f is the focal length of the lens. Always draw light going from left to right. Object distances to the left and image distances to the right are positive. Object distances to the right (light converging on the lens) and image distances to the left (light diverging from the lens) are negative. Negative distances are called virtual. Focal length f is positive for convex lenses and negative for concave lenses. If any two quantities are given, the third may be calculated. The ratio v/u equals the magnification.

Using the lens equation we find that if an object is placed 25 cm in front of a convex lens whose focal length is 10 cm, then the image distance works out at +16.7 cm. This example is illustrated in the first ray diagram on page 146. The image is real and could be focused on a screen. Alternatively, if the same object is placed 7 cm in front of the same lens, the image distance works out at −23.3 cm. The image is virtual. It cannot be focused on a screen and it can only be seen by looking through the lens. In this second example, illustrated by the second ray diagram, the lens is used as a magnifying glass.

lens equation

Drawing ray diagrams to scale is an alternative to using the lens equation. The sequence of events is as follows:

- draw a straight line from side to side across the page. This line is called the principal axis.
- draw a line at right angles to the principal axis to represent the lens.
- mark the positions of the two focal points at equal distances from the lens but on opposite sides. Use a suitable scale for the focal length.
- draw a line at right angles to the principal axis and to the left of the lens to represent the object. Use the same scale for object distance as you did for focal length.
- from any off-axial point on the object line, draw a straight line through the point where the lens line crosses the principal axis.
- draw a straight line from the same point on the object parallel to the principal axis. From the point where this line crosses the lens line, draw a straight line through the focal point on the right-hand side of the lens.
- these last two lines cross at the image. Draw the image and measure the image distance.

The broken lines in the second diagram indicate a virtual image through which no light has passed.

A lens with short focal length bends the light more than a lens with long focal length. The reciprocal of the focal length f, measured in metres, is called the power P of the lens. P is measured in dioptres.

$$P = \frac{1}{f}$$

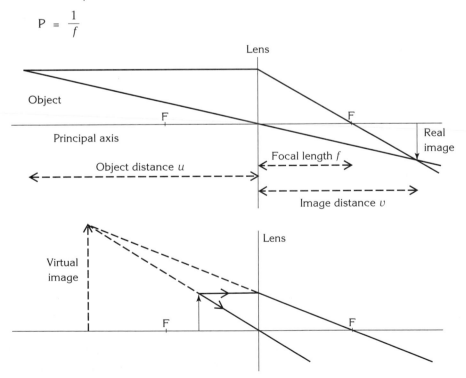

Lenz's law of electromagnetic induction: See *electromagnetic induction, electromagnetic induction (laws of)*.

leptons are those fundamental particles that do not respond to the strong interaction. There are 12 of them altogether. Six leptons are shown in the following table; the other six are their antiparticles.

Generation	Particles of charge −e		Particles of charge zero	
	Particle	Rest mass	Particle	Rest mass
First	electron, e	0.51 MeV/c²	electron neutrino, v_e	0
Second	muon, μ	106 MeV/c²	muon neutrino, v_μ	0
Third	tau, τ	1784 MeV/c²	tau neutrino, v_τ	0

There are three principal leptons which carry negative charge and mass: the electron (e⁻), the muon (μ⁻) and the tau (τ⁻). These, and the corresponding positively charged antiparticles, are subject to gravity, to electromagnetic fields and to the weak interaction. To each of these six fundamental particles there corresponds a neutrino or an antineutrino, giving us 12 leptons in all. The six neutrinos and antineutrinos carry no charge and any mass they might have is too small to measure. They are immune to the electromagnetic interaction and to gravity, they are subject to the weak interaction only. Any nuclear event is controlled by the weak interaction if it in some way involves neutrinos or antineutrinos.

The 12 leptons are separated into three generations. Each generation starts with one of the three negatively charged leptons and includes its antiparticle and the corresponding neutrino and antineutrino.

Why don't we see more of them? The electron is stable but any positrons lucky enough to be around are soon annihilated. The heavier muon (about a tenth of the proton mass) decays after an average lifetime of about 2.2 microseconds into an electron, a neutrino and an antineutrino. The tau particle is not far short of two proton masses and decays in a number of ways to produce muons, electrons or even pi mesons. Muons and pi mesons end up as electrons. Only neutrinos in the first generation (from the Sun etc.) can exist in any numbers and they are largely inactive. The material universe is mostly protons, neutrons, electrons and neutrinos, We live in a low energy universe.

Apart from everyday electron behaviour, beta decay processes and positron annihilation, lepton interactions are observed only in high energy collision experiments. A number of conservation laws apply:

- mass–energy
- charge (either e⁻, e⁺ or zero)
- spin (all twelve leptons carry spin ½)
- lepton number (conserved separately for each generation – +1 for negative leptons and their neutrinos, −1 for the other six).

The neutrino idea was introduced into physics in order to reconcile beta decay with the concept of mass–energy conservation. The electron neutrino (so-called at the time) was later discovered to be the electron antineutrino.

lift is the upward thrust on an *aeroplane* brought about by the different airflow speeds over the upper and lower surfaces of the wings. It is explained in terms of *Bernouilli's principle*.

light as a transverse wave motion accounts for the phenomenon of *polarisation*. If light were a longitudinal wave motion (like sound) there would be no polarisation effects. (See *transverse waves, longitudinal waves.*)

light-dependent resistor (l.d.r.): a resistor whose resistance decreases if the ambient light level increases. The current I in a conductor of cross-sectional area A is given by the relationship:

$$I = nAqv$$

where there are n charge carriers per unit volume moving with an average drift velocity v each carrying charge q. Thin layers of semiconducting material can be constructed with the special property that the number of charge carriers per unit volume of semiconductor, n, increases with ambient light level. There is then more current for a given potential difference across the specimen so resistance has fallen.

light-emitting diodes (l.e.d.) are *semiconducting diodes* which emit light when they are forward biased and carry current. A light emitting diode is made from semiconducting materials which requires a forward bias of 2 V or so before forward current rises appreciably. A proportion of the electrons flowing across the junction into the p-type material recombine with an equal number of holes moving the other way. The energy released upon each recombination appears as a single *photon*. The colour of the light emitted depends on the forward voltage at the 'knee' of the current–voltage characteristic.

light rays are the straight lines light is imagined to travel along when approaching a mirror or a refracting surface. Light is real but light rays are imaginary. Light rays are fundamental to the principle of rectilinear propagation, the principle that light travels in straight lines. (See *ray diagram.*)

limit of proportionality is the point on a *stress–strain curve* at which the line stops being straight. It marks the maximum stress for which *Hooke's law* applies.

limiting value is the value of a physical quantity calculated from the gradient of a graph. A good example is average speed and instantaneous speed. The average speed during the time interval $(t_B - t_A)$ for the distance–time graph shown in the diagram below is BC/AC. Imagine this time interval shrinks slowly, with point A remaining fixed but point B sliding down the curve towards A. Point B also gets ever closer to the tangent at A. In the limit, when B reaches A, the line AB coincides with the tangent to the curve at A and the average speed is identical to the instantaneous speed. The value of BC/AC is identical to the gradient of the tangent at A.

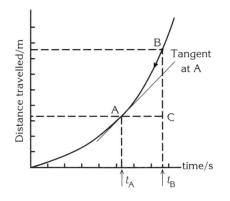

This idea is summarised in the statement: 'the instantaneous speed at time t_A is the limiting value of the average speed calculated for a very short time interval beginning at time t_A.'

linear momentum is defined by the word equation:

linear momentum = mass × velocity

It is a *vector quantity* and has the same direction as its velocity vector component. Mass, a *scalar quantity*, acts like a multiplier. The *conservation of linear momentum* is perhaps the most fundamental law of Newtonian mechanics and the concept of force draws its definition from rate of change of linear momentum. The SI unit for linear momentum, the newton second, comes from the definition of impulse.

(See also *impulse*, *Newton's laws of motion*.)

lines of force: see either *field lines (electric)* or *field lines (magnetic)*.

liquid pressure (laws of): pressure p, in a stationary liquid mass, is calculated from the following equation:

$$p = p_0 + h\rho g$$

where p_0 is the pressure on the surface, h is the depth below the surface, ρ is the density of the liquid and g is the *acceleration of free fall*. (Gases do not obey this simple form of the law: they are compressible so density increases with depth.) The pressure within a liquid mass:

- acts at every point in the liquid mass, not just at the surfaces
- increases linearly with the depth below the surface
- increases linearly with the density of the liquid
- increases linearly with the gravitational field strength
- is independent of the shape of the container
- acts in any direction at a point
- exerts a normal thrust on any solid surface with which it is in contact.

liquids: fluids in which every molecule is in contact with its neighbouring molecules but is not attached to any of them. The short-range order which accounts for crystal structure in solids, especially when it continues into the long range, is absent in liquids. Molecules in a liquid are not tied together; they are free to move about within the body of the liquid but they are not free to separate. As a consequence, a liquid is almost incompressible but has no shape of its own. A liquid takes the shape of its container. Like gases, liquids can transmit pressure but not force. The molecular speeds have a spread of values and this is linked to the process of evaporation.

There is no confusing a solid with a gas but the liquid and solid state sometimes overlap. The reason is that molecules in a liquid are in contact with one another giving weak intermolecular forces (such as *Van der Waals forces*) a chance to operate. The larger the molecule and cooler the liquid mass, the greater the probability that these forces will be active. Substances such as glass, toffee and pitch are brittle at low temperatures and 'runny' at higher temperatures without a definite transition temperature at which the substance melts. Even when cold and hard, these substances will appear hard to fast-acting forces (they either shatter or bounce) but respond in the same way as liquids to slow-acting forces (they flow, deform and collapse). Even so, there is still value in the general distinction between *solids* which answer to forces and *fluids* which answer to pressures. (See *laws of liquid pressure, pressure, thrust, Bernouilli's principle, Archimedes' principle, solid, gas, viscosity, surface tension, evaporation and cooling.*)

localised vector: a vector which has magnitude and direction and which acts through a particular point. (See *force, a localised vector.*)

longitudinal waves are *progressive waves* whose displacements are along the line of energy flow. *Sound waves* are longitudinal waves. (See also *stationary waves, transverse waves.*)

loudness is the subjective estimate of sound intensity made by the ears. However, experiments with large numbers of people judging the loudnesses of sounds of known intensity, support the following relationship between loudness L and intensity I:

$$L = 10 \log_{10} \frac{I}{I_0}$$

where L is the loudness of the sound in decibels, I is the sound intensity in W m^{-2} and I_0 is the minimum sound intensity which is just audible to a 'standard ear'. An accepted value of I_0 is 1×10^{-12} W m^{-2} and 120 decibels corresponds to a sound which is so loud that it begins to hurt. The intensity value for this level of loudness is 1 W m^{-2}. Ears then operate across twelve orders of magnitude – an astonishing feat.

luminosity L of a star is the total radiant power output in watts. If luminosity and surface temperature are both known, then using *Stefan's law*, the radius of the star can be found:

$$L = A\sigma T^4 = 4\pi r^2. \sigma T^4$$

See *star.*

What other subjects are you studying?

A–Zs cover 18 different subjects. See the inside back cover for a list of all the titles in the series and how to order.

magnet: a block of permanently magnetised ferromagnetic material, usually with opposite poles at the far ends. (See *ferromagentism*.)

magnetic effect of current: the power of electric current to create a magnetic field in its immediate vicinity. The diagram below shows three cases: a straight wire, a ring and a solenoid. The *right-hand grip rule* gives the direction of the magnetic field. For a straight wire, the thumb points along the current direction and the fingers follow the direction of the circular magnetic field lines. For the solenoid, the fingers follow the current direction and the thumb points along the solenoid towards the north-seeking pole. The direction of the magnetic field near the ring can be worked out from either of these applications.

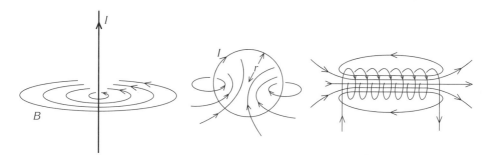

The magnetic field strength B a perpendicular distance r from a long straight conductor carrying current I in vacuum is given by:

$$B = \frac{\mu_0 I}{2\pi r}$$

The magnetic field strength B at the centre of a circular ring of radius r and with N turns and carrying current I in vacuum is given by:

$$B = \frac{\mu_0 N I}{2\pi r}$$

The magnetic field strength B along the centre of a long solenoid with n turns per metre of length and carrying current I in vacuum is given by:

$$B = \mu_0 n I$$

Always draw at least four or five magnetic field lines and remember to leave gaps in the lines and coils as shown in the diagram. If these gaps are absent, the diagrams become ambiguous and you will lose marks.

magnetic field strength, B, at a point is defined by the equation

$$F = BIl$$

where F is the force on a straight conductor of length l which carries current I. The value of F is in newtons, I in amperes, B is in tesla (T), the SI unit of magnetic field strength and l is in metres. This equation can be used to define B because F, I and l can all be measured independently and separately from B. (See *magnetic effect of current, force on moving charge, magnetic flux*.)

magnetic flux, ϕ, threading an area is the product of *magnetic field strength* and the component $A \sin \theta$ of the area A at right angles to the magnetic field. The word equation is:

magnetic flux ϕ = magnetic field strength $B \times$ area $A \times \sin \theta$

Where θ is the angle between the area and the magnetic field direction (see diagram below):

$$\phi = B \times A \times \sin \theta$$

The SI unit for magnetic flux is the weber (Wb):

weber (Wb) = tesla (T) $\times$ metres squared (m²)

The definition of magnetic flux should be clear from the following diagram:

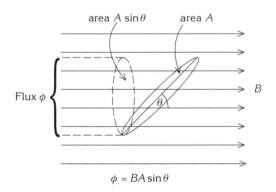

$$\phi = BA \sin \theta$$

Magnetic field strength is sometimes called magnetic flux density.

magnetic flux density, B, is another name for *magnetic field strength*. Lines of magnetic force or flux are used to describe the shape of a magnetic field near a magnet or solenoid. The density of the *lines of force* increases in proportion to field strength. The number of lines of force can be defined in such a way that magnetic flux density is identical to the magnetic field strength.

magnetic flux linkage (See *flux linkage*.)

magnitude: see *star*.

malleability is the extent to which a material can be hammered out into a thin, flat sheet. It is a characteristic feature of metals (hot more than cold) and especially of gold. Formally, it is the ability to undergo *plastic deformation* in response to a compressive load.

manometer: a device based on a U-tube for measuring pressure difference. The principle is shown in the diagram on page 153. The pressure difference between the inside of the vessel and the outside is calculated from the equation:

$$p = p_0 + h\rho g$$

where p is the pressure inside the vessel, p_0 is the pressure outside the vessel, h is the difference in level between the two surfaces in the U-tube, ρ is the density of the liquid and g is the acceleration of free fall.

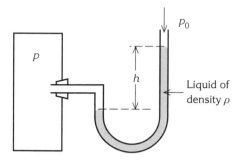

p

p_0

h

Liquid of
density ρ

mass is so basic a quantity in Newtonian mechanics that it defies definition in terms of simpler concepts. This is why we have to make do with less-than-satisfactory descriptive phrases like 'mass is the amount of matter in a body'. Mass is a *scalar quantity*. Its unit, a *base unit* in SI, is the kilogram (kg). Mass is measured, in principle, with the chemical balance or its equivalent. It cannot have a negative value.

Mass is distinct from amount of matter measured in *moles*. A number of moles is how many identical, countable bits are present. The two concepts are related through the idea of *molar mass*.

For the difference between gravitational mass and inertial mass, see *acceleration of free-fall*.

mass defect, written Δm, is the difference in mass between a nucleus and the combined masses of its constituent neutrons and protons. $\Delta m c^2$ is the energy that would be needed to pull all the neutrons and protons out of the nucleus. This energy is called the *binding energy*.

mass–energy equivalence is the idea expressed in the equation:

$$E = \Delta m c^2$$

If a body acquires extra kinetic energy E, then its mass increases by the amount Δm given in this equation. The word 'equivalence' is important. During nuclear fission, mass is not lost; it is dispersed as an equivalent amount of energy. There is no such thing as mass–energy conversion. (See *binding energy*.)

mass number, A, is the number of nucleons (neutrons and protons) in a nucleus. The definition can be stated as a simple addition; mass number equals *atomic number Z* plus *neutron number N*:

$$A = Z + N$$

mass spectrometer: an instrument in which a beam of positive ions is projected into a plane at right angles to a strong, uniform magnetic field. The ions follow a semicircular path to a photographic plate which forms an accurate record of the point of arrival. The radius of the semicircle is measured, the magnetic field strength is known so the mass of the positive ion can be calculated from the relation $r = mv/Bq$.

A
B
C
D
E
F
G
H
I
J
K
L
M
N
O
P
Q
R
S
T
U
V
W
X
Y
Z

153

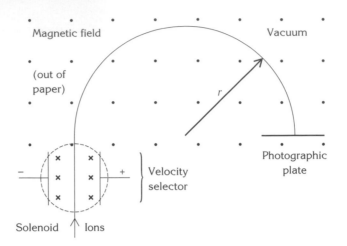

The positive ion speed v is found as follows. The source of positive ions includes a velocity selector which consists of an electric field E and a magnetic field B' at right angles to one another and at right angles to the ion beam. Only those ions whose speeds are E/B' can get through. See *electron beams*.

Maxwell's corkscrew rule: a rule for finding the direction of the magnetic field near a straight wire. Imagine you are driving a right-handed corkscrew in the direction of the current in the wire. The direction in which you turn the handle of the corkscrew is the direction of the circular magnetic lines of force. (See *right hand grip rule.*)

mean power in resistive load is the average power dissipated in the load during one complete cycle by a sinusoidal AC current. The circuit in the diagram on page 155 links an AC power supply V_0 sin ωt to a resistor R. If V_0 is 6 V and R is 15 Ω, then I_0 is 0.4 A.

The instantaneous AC current is linked to the instantaneous voltage by *Ohm's law*:

$$I = I_0 \sin \omega t = \frac{V_0}{R} \sin \omega t$$

The instantaneous power P equals IV and we can write:

$$P = IV = I_0 V_0 \sin^2 \omega t$$

It is clear from the lower graph (on page 155) that the average value of $\sin^2 \omega t$ is 0.5 exactly. The graph is symmetrical above and below the 0.5 line. So <P>, the mean value of the power supplied to the resistor, is one half the peak value:

$$<P> = \tfrac{1}{2} I_0 V_0 = \tfrac{1}{2} I_0^2 R = \frac{V_0^2}{2R}$$

Compare this result with the DC case:

$$P = IV = I^2 R = \frac{V^2}{R}$$

If P equals <P>, then I equals I_{rms} and V equals V_{rms}. Comparing the two sets of equations gives:

$$I_{rms}^2 = \tfrac{1}{2}I_0^2 \quad \text{or} \quad I_{rms} = \frac{I_0}{\sqrt{2}}$$

$$\text{and} \quad V_{rms}^2 = \tfrac{1}{2}V_0^2 \quad \text{or} \quad V_{rms} = \frac{V_0}{\sqrt{2}}$$

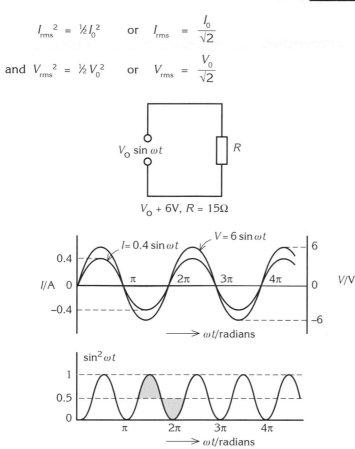

$V_0 + 6V$, $R = 15\Omega$

(See also *r.m.s. values.*)

mean square speed $<c^2>$ is defined by the equation:

$$<c^2> = [c_1^2 + c_2^2 + c_3^2 \dots c_N^2]/N$$

where $c_1, c_2, c_3 \dots c_N$ are the instantaneous speeds of the N molecules in a gas sample. Mean square speed $<c^2>$ is the average value of (molecular speed ^{2}).

Although $<c^2>$ is defined by the above equation, it is never calculated using this equation. Once defined, its properties can be determined and its most important property is the link it provides between gas pressure and temperature. (See *kinetic theory of gases.*)

measurement is the process of linking a physical quantity with a magnitude. There are three stages in the complete process: the physical quantity has to be defined, instrumentation must be provided and the unit must be known. An instrument is a device which compares two magnitudes. A ruler compares two lengths, a chemical balance and a box of weights will compare two masses and a galvanometer will compare two voltages. The output of the instrument usually appears on a scale of some sort. This scale can be *calibrated* in terms of the unit by noting the outputs for a range of standard inputs.

mercury barometer: an instrument which measures atmospheric pressure by balancing it against the pressure at the base of a column of mercury. The device shown in the diagram below is sometimes called a mercury barometer, sometimes a 'simple barometer'. The pressure of the atmosphere p_0, equals the pressure at the bottom of the mercury column at the same level as the surface of the mercury in the reservoir. Since there is a vacuum (and therefore zero pressure) above the mercury, the pressure at the bottom of the mercury column is Hpg where H is the vertical distance from the surface of the mercury in the bowl to the top of the mercury meniscus in the tube, p is the density of mercury and g is the acceleration of free fall. We have, then:

$$p_0 = Hpg$$

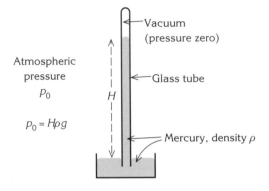

The Fortin barometer is an accurate version of the simple barometer. It has an arrangement for setting the mercury at just the right level in the reservoir and a vernier scale for measuring the height of the mercury column. There is also a thermometer should a temperature correction for the density of mercury be needed.

mercury-in-glass thermometer: a well-known liquid-in-glass thermometer in which mercury stored in a glass bulb expands along a uniform capillary in the glass stem as the temperature rises. The mercury always expands to the same point for a given temperature. The stem is usually calibrated to read temperatures between 0°C and 100°C, but the instrument can be constructed to cover other ranges. Mercury thermometers are usually calibrated by marking the glass at the 0°C mark using iced water and at the 100°C mark using steam over boiling water at normal pressure. A scale with a hundred equal divisions is then marked on the stem starting at 0°C and finishing at 100°C. This method of calibration assumes that the temperature θ recorded by the thermometer is related to the mercury column length in the following manner:

$$\theta = \frac{l_\theta - l_0}{l_{100} - l_0} \times 100$$

In formal language, this equation defines temperature on the mercury-in-glass temperature scale.

Care should be taken when using a mercury thermometer to avoid systematic errors. The instrument should be used upright (hold it upside down and watch the temperature rise a degree or so) and dipped in the liquid up to the immersion mark on the thermometer. Solids must be brought into good thermal contact with the thermometer and gas temperatures can only be measured if the thermometer is given plenty of time to settle. The mercury thermometer cannot be used to measure temperatures of small objects.

mesons are made up of a quark and an antiquark with opposite values of the same colour. They can have integral or zero charge, baryon number zero and spin 1 or 0. The π minus for instance is dū. Four such π mesons or pions can be formed from first generation quarks and are created frequently in collision experiments. One consequence of the quark–antiquark arrangement is that mesons can annihilate in interactions. Meson number, unlike baryon number, is not necessarily conserved in an interaction event. Mesons decay into leptons (and eventually into electrons and antineutrinos) and photons. The table shows details of those mesons that were discovered in the 1940s. Pi mesons are sometimes called pions and K mesons are sometimes called kaons.

Name	Quantum numbers			Quark content
	Q	B	S	
π meson				
neutral	0	0	0	see below*
positive	+1	0	0	u anti-d
negative	−1	0	0	d anti-u
K meson				
neutral	0	0	+1	d anti-s
positive	+1	0	+1	u anti-s
negative	−1	0	+1	s anti-u

* the neutral π meson occurs as a mixture of (u anti-u) and (d anti-d) particles.

metallic bond: a form of attachment between atoms in which each atom releases either one or two electrons to a collective 'electron gas' and the resulting positive ions build up a granular structure, kept stable by these electrons which inhabit the spaces between the atoms. The strength of the bond accounts for the hardness and high melting point of many metals. The presence of mobile electrons accounts for the electrical and thermal conduction properties of such solids.

metals are those substances which, in the solid state, consist of a latticework of positive ions held in a stable condition by free electrons. The electrons behave like a gas which inhabits the open spaces within the crystal lattice. The number of electrons is just sufficient for their combined negative charge to cancel the net positive charge of the lattice. The directional quality of the inter-atomic forces is not strong enough to build up and hold together large crystal structures. Because of this, metals consists of numerous small crystals or grains. The sizes of these grains and the forces locking them together decide the particular physical properties. The existence of these tiny grains also accounts for the possibility of beating metals out into thin sheets (malleability), the readiness with which they can be drawn out into wire form (ductility) and the stability and sharpness of ground edges. In contrast, the strengths of these inter-atomic forces, leaving aside their modest directional qualities, account for the high values of Young's modulus, hardness, great strength and high melting point of many metals. (See *metallic*.)

metre: a *base unit* in *SI*. It is the SI unit of length and has the symbol m. The metre equals the distance travelled in vacuum by light during a time interval of $1/c$ seconds, where c is the *speed of light* in vacuum. The original historical definition of the metre was the length of the

Earth's equator divided by 40 million, This implies a value of 6366 km for the radius of the Earth at the equator compared with the present estimate of 6378 km. It is now defined as a given number of wavelengths of krypton–86 in vacuum.

MeV is an energy unit. It is an abbreviation for a million *electron volts*. It equals 1.602×10^{-13} J.

MeV/c² is a mass unit. It is the mass equivalent of 1 MeV. It equals 1.78×10^{-30} kg. Proton mass can be written 940 MeV/c^2 and similarly electron mass 0.512 MeV/c^2.

microwaves are *electromagnetic waves* with wavelength in the region from about 1 mm to about 10 cm. A typical microwave source is either a low power klystron valve (a few mW) or a high power magnetron valve (up to a few kW). The microwaves are radiated from electrons oscillating in an aerial or within a cavity. The same waveband is used for radar.

Microwave wavelengths are just short enough to allow modest focusing and approximately straight line transmission. This is important in the communication field because it cuts down the power needed to send information from one parabolic transmitter to the next parabolic receiver. Microwave frequencies are in the range 300 GHz (300×10^9 Hz) to 3.0 GHz (3.0×10^9 Hz). These high frequencies result in a high capacity for carrying information and account for the emphasis placed on this waveband by communication engineers. This application is now giving way, in part, to optical fibres with their much higher information capacities.

Much information about the Earth's surface can be collected on microwave frequencies. This explains the emphasis on these wavelengths in the design of the European survey satellites ERS1 and ERS2.

Millikan's oil drop experiment was the first tolerably accurate and direct measurement of the charge on an electron. The experiment established that electric charge is a quantised physical variable with a minimum increase or decrease in value now identified as the charge on one electron.

The apparatus, in broad outline, consisted of two parallel brass plates. The upper plate had a hole in the centre, the plates were mounted exactly horizontal and the voltage between them could be adjusted to any value from zero to about 6 kV, positive or negative. A microscope was used to watch oil drops from an atomiser falling vertically through the hole. The oil drops were illuminated from the side and charged by friction at the exit of the atomiser.

The first stage of the experiment involved trapping a charged oil drop by adjusting the electric field so that the upwards force equalled the downwards force. The forces on an oil drop are the weight, mg, the buoyancy (which will be ignored) and the electric force qV/d. In these circumstances:

$$mg = qV/d$$

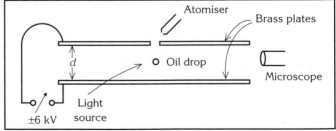

Casing to prevent draughts

The second part of the experiment involved switching off the electric field and measuring the terminal speed of the oil drop. In these circumstances (again, ignoring the buoyancy) the viscous drag $6\pi a\eta v$ equals the weight mg. The viscous drag force is given by the expression $6\pi a\eta v$ (*Stokes' law*) where a is the radius of the oil drop, η the viscosity of air and v the terminal speed of the oil drop. We have:

$$mg = 6\pi a\eta v$$

The second diagram (below) shows the *free-body force diagrams* for these two situations.

Knowing the density of the oil it is possible to eliminate the radius of the oil drop from these two equations and thereby derive a single equation with q as the single unknown quantity.

Millikan measured the charges on many oil drops and found that the measured value of q was always a small integral multiple of -1.59×10^{-19} coulombs. He took this to be the charge on one electron.

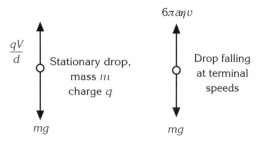

models of the universe set out to explain in terms of a few basic principles the complex patterns of behaviour observed among the stars.

Greek astronomy took as its basis the sky as a permanently patterned background of stars which rotated about the Earth's polar axis once a day. This patterned background enabled the changing positions of the sun, the moon and the five known planets to be observed and recorded. They tried to explain these movements in terms of many uniform circular motions compounded together. The idea that motion along a circular path at constant speed is the one perfect motion which needs nothing to sustain it, was the basic model for Greek cosmology. This model had the merit of encouraging the orderly collection of data over many centuries. The down-side was the ever-increasing complexity of the system – almost 150 separate circular motions to account for the varied positions of the sun, the moon and five planets.

Eventually, Copernicus showed (1543) that the number of necessary uniform circular motions would be more than halved by assuming that the Sun, not the Earth, was the central body. Then Kepler (1609 and 1618) rejected countless uniform circular motions in favour of a few elliptical orbits – one for each planet and with the Sun at one focus of each ellipse. Galileo explained why, if the Earth was rushing round the Sun at enormous speed, the air would move with us and we wouldn't notice we were moving.

Gravitational astronomy began with a derivation of *Kepler's laws* from *Newton's law of universal gravitation* by Sir Isaac Newton (1687). This derivation would not have been possible without the creation of Newtonian mechanics and would have been infinitely more difficult without his invention and development of the differential calculus. Gravitational astronomy gave an extraordinarily accurate account of the motions of the planets and their moons, and of the shape of the Earth and its tides.

Uranus, discovered by William Herschel in 1781, was painstakingly tracked over the next few decades and its orbit was found to be far from elliptical. Two men, Adams in England and Le Verrier in France, independently calculated the position of an eighth planet close enough to Uranus to account for the distorted orbit. Neptune's discovery in 1846, close to its calculated position, is recognised as the high point of gravitational astronomy.

Modern astronomy involves the use of atomic and nuclear physics to analyse and understand the make-up of stars and galaxies and the universe they inhabit. The prevailing model involves about 10^{13} galaxies, each containing 10^{11} stars (some a lot more and some a lot less) and all of them spreading outwards as part of an expanding universe. This simple picture understates the complexity and variety of objects within the universe.

moderator: a large mass of graphite which provides the matrix in which the *fuel rods*, the coolant pipes and the *control rods* of a nuclear power plant are supported. The principal function of the moderator is to slow down to thermal energies the fast neutrons emitted in the nuclear *fission* reactions in the control rods so that further fission reactions can proceed.

modulation originally meant the combination (in a form of addition process) of an information wave with a carrier wave before transmission. An example would be the mounting of a sound wave signal on a much higher frequency radio-wave carrier. The opposite procedure, called demodulation, took place after signal reception.

The word modulation now describes any signal modification process which prepares a signal for transmission. See *amplitude modulation*, *frequency modulation* and *pulse code modulation*.

molar heat capacities of gases are the heat capacities per *mole* for a gas. Two values are important: the molar heat capacity at constant volume and the molar heat capacity at constant pressure. *Specific heat capacity* (heat capacity per unit mass) works well with solids or liquids. For gases, 'specific' becomes 'molar' and capacity becomes plural.

Why the plural form 'capacities'? The reason is linked to the ease with which gases change volume and is derived from the *first law of thermodynamics*. The first law of thermodynamics states that any increase in the internal energy of a system, ΔU, equals the sum of the heat flowing into the system, ΔQ, and work done on the system ΔW. If the change takes place at constant pressure, then ΔW equals $-p \Delta V$ and we have:

$$\Delta U = \Delta Q - p\Delta V$$

Remember that ΔW is positive if work is done on the system. ΔV is then a decrease in volume and it is negative. The product $(-p \Delta V)$ contains two negative quantities and is, therefore, a positive quantity.

Specific heat capacity is the heat transferred per kilogram of substance per degree kelvin temperature rise. It is the value of ΔQ. But temperature rise is related to ΔU. Solids and liquids are almost incompressible, ΔV is effectively zero, ΔW is zero and ΔU equals ΔQ. But for gases, ΔV is significant. The heat needed to give the increase in U, ΔU, corresponding to a one degree kelvin temperature rise depends on what proportion of ΔQ is siphoned off to provide the energy $p \Delta V$ needed if the gas is expanding and working on the surroundings. There is room for an infinite number of specific heat capacities for a gas. Fortunately, any change in the state of a fixed mass of gas can be represented as the sum of two changes, one at constant volume and one at constant pressure. So we only need two specific heat capacities.

Now for the word 'molar'. Intermolecular forces in a gas are so small as to be negligible. Internal energy is the kinetic energy of the molecules. But the average molecular kinetic energy in a gas is proportional to absolute temperature. It follows that the increase in internal energy per unit temperature rise in a gas depends only on the number of molecules present and not directly on the mass of gas present. The heat capacities per mole are the same for all gases.

Taking these two ideas together, the physical quantities which matter are the molar heat capacity at constant volume and the molar heat capacity at constant pressure.

molar mass: the mass of one *mole* of substance. It varies from one substance to another according to the mass of the relevant entity whose count identifies the mole.

mole: the SI base unit for amount of substance. The symbol is mol. It is the amount of substance that contains as many elementary entities as there are in 0.012 kg of carbon-12. See *Avogadro constant*.

moment of a force about a point is the product of the force and the perpendicular distance from the point to the line of action of the force. The unit is the N m. Do not confuse this with the definition of the *joule*. Energy is a *scalar quantity* and the distance measured in the definition of the joule is parallel to the direction of the force. In the N m as a unit of moment of a force, the distance and the force are at right angles. Moment of a force is not a scalar quantity, it is a *pseudo-vector* but this does not matter at A level.

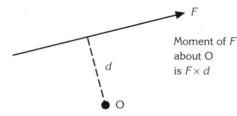

Moment of F about O is $F \times d$

There are two ways forward. The sum of the moments of several forces acting on a body may all cancel in which case the body will be held in an equilibrium state (dealt with in the entry *principle of moments*). The alternative is for the sum of the moments to add up to a resultant turning force; this case is dealt with under *torque*.

(See also *principle of moments, couple, forces in equilibrium*.)

moment of inertia is defined by the word equation:

$$\text{moment of inertia} = \frac{\text{torque}}{\text{angular acceleration}}$$

$$I = \frac{T}{\alpha}$$

The unit of moment of inertia is kilogram metre^2 (kg m^2). Moment of inertia plays an analogous part in *rotational motion* to mass in *translational motion*. The moment of inertia of a disc-shaped object of given radius, such as the flywheel in a car engine, about an axis normal to the disc and passing through its centre, increases if the mass distribution can be moved towards the edges. A heavy door has a much higher moment of inertia than a light door. It is much harder to get moving and once started it is harder to stop again.

momentum: see *linear momentum, angular momentum*.

motor boats are driven through the water by the reaction of the water to the thrust developed by a spinning propeller parallel to the keel. If the rudder is inclined to the keel, the faster flow of water over the outer edge of the rudder compared with the slower current over the near edge gives, according to the *Bernouilli principle*, a sideways thrust on the rudder which turns the boat in the water. This sideways thrust only operates when the boat is driven through the water by the propeller. (See also *sailing boats*.)

motor effect: the force experienced by a current-carrying conductor in a magnetic field. If *B* is the strength of the (uniform) magnetic field, *l* is the length of conductor under consideration and *I* is the current in the conductor, then the force *F* on the conductor is given by:

$$F = BIl \sin \theta$$

where θ is the angle between the current direction and the *magnetic field strength*. The direction of the force is given by *Fleming's left hand rule*.

MRI: an acronym for magnetic resonance imaging. The imaging system is built around the behaviour of atomic nuclei, especially hydrogen, in powerful magnetic fields in excess of 1 tesla. The magnetic resonance image of the human body is built up in stages:

● First, the body is placed parallel to and within a powerful uniform magnetic field, set up by a high and permanent current in a superconducting solenoid. Next, a second coil imposes a second magnetic field in the same direction as the first but strong at one end, weak at the other and diminishing uniformly along the body. Each slice of the body has its own magnetic field strength.

● The human body is rich in hydrogen atoms. Most hydrogen nuclei precess parallel to the strong magnetic field direction but some precess in the anti-parallel direction. The precession frequency is proportional to the magentic field strength *B*.

● A brief signal from a radio frequency coil excites the hydrogen atoms within one slice of the human body where the magnetic field is just the right strength for that radio frequency. All hydrogen atoms within the slice and precessing in the parallel direction flip to the anti-parallel state.

● Once the radio signal has passed, the hydrogen atoms revert to their previous states and release the absorbed energy as a radio signal. Each point in the plane is scanned in turn and the reception of the full set of signals from any one plane gives the distribution of hydrogen atom concentration within that plane. Hydrogen atom concentration is an indicator of tissue type. An image of the slice can be built up.

● The scanner then moves on to the next slice.

mutual inductance, *M*, of two circuits equals the *e.m.f.* induced in one of the circuits when the current in the other circuit changes at the rate of one ampere per second. The word equation definition is:

$$\text{mutual inductance} = \frac{\text{e.m.f. induced in one circuit}}{\text{change of current per second in the second circuit}}$$

The numerical value of the mutual inductance can be shown to be the same whichever of the two circuits is chosen for the changing current. The SI unit for mutual inductance is the henry (H). (See also *self-inductance*.)

neutral point: a point within a magnetic field where the resultant *magnetic field strength* is zero. Neutral points are especially significant when they result from the overlapping of a non-uniform magnetic field (like that near a bar magnet) and a uniform magnetic field (like the horizontal component of the Earth's magnetic field on a local scale). They occur at those points where the magnetic field strengths of the two magnetic fields are equal and opposite.

Since neutral points occur where the two magnetic field strengths are equal and opposite, if one field is known then so is the other. For instance, if the current in the vertical wire is known and if the distance of the neutral point from the wire is measured, then the strength of the magnetic field set up by the current in the wire at the neutral point can be calculated. The horizontal component of the Earth's magnetic field strength at the point in question is then known.

neutron: an elementary particle which carries no electric charge and with a mass almost the same as that of a hydrogen atom.

neutron number N is the number of neutrons in a nucleus. (See also *atomic number*.)

neutron star: the *end state* of a main sequence star whose mass is between eight and 20 solar masses. The stellar remnant following a supernova explosion has a mass greater than the solar masses but a radius of roughly 10 km. This corresponds to a density of 10^{16} kg m^{-3} and upwards. This high density is linked to the conversion of core protons and electrons (under the weak interaction) into neutrons, together with the loss of the electron degeneracy pressure which supports white dwarfs. They are observed as pulsating radio sources (*pulsars*) of which several hundred are known. Neutron stars are identified also with binary X-ray sources.

The flashes of radio waves, which are the characteristic property of pulsars, are attributed to a steady narrow-beam radio emission coupled with a rapid period of rotation. This period, ranging from a few milliseconds to a few seconds, is explained by the conserved angular momentum and the much reduced moment of inertia of the contracting core. For any one star, the period is stable to about one part in 10^{13} and this matches the stability of atomic clocks. See *end states of stars*, *four fundamental interactions*.

newton is a short way of writing kilogram metre per second squared (kg m s^{-2}). It is the SI unit of force. The statement 'One newton is that force which is just strong enough to give a one kilogram mass an acceleration of one metre per second squared in the direction of the force' is true. But it defines a force of *one* newton, not the unit itself, *the* newton. What you should write depends on what you are asked. The word equation definitions of force and the newton are as follows:

$$\text{force} = \text{mass} \times \text{acceleration}$$

$$\text{newton} = \text{kilogram metre per second squared (kg m s}^{-2})$$

(See also force, Newton's second law of motion.)

Newton's first law of motion

'A body subject to zero resultant force either keeps still or moves with constant velocity.'

This law is a kind of datum-line. If the behaviour of the body under zero resultant force is given by the first law and if the behaviour of the body under non-zero resultant force is given by the second law, then the difference between the happenings in the two laws gives the effect of the non-zero resultant force. See Newton's laws of motion, Newton's second law of motion, Newton's third law of motion.

Newton's law of cooling states that the rate at which a body cools is proportional to the temperature excess of the body above the surroundings. This is not a universal law. It applies provided (i) the body is in a strong draught and (ii) the temperature excess is not greater than about 30 K. If there is no strong draught, convection cooling becomes progressively less effective and the cooling rate diminishes unacceptably. If the temperature excess rises much above 30 K, the radiation loss is too rapid and this interferes with the law. Notice that the law satisfies the condition for exponential decay.

Newton's law of universal gravitation states that there is an attractive gravitational force F between any two particles of matter in the universe which is proportional to the product of their masses and inversely proportional to the distance between them.

If m_1 and m_2 are their masses and if r is their distance apart, then the gravitational force of attraction, F, is given by:

$$F = G\frac{m_1 m_2}{r^2}$$

where G is the universal gravitational constant with a value 6.67×10^{-11} N m^2 kg^{-2}.

A law which applies only to particles would be of very limited application, but Newton was able to show that, as far as its external gravitational field is concerned, a uniform sphere (such as the Earth, Moon or Sun) behaves as if its whole mass were concentrated at its centre.

Worked example

The masses of the Sun, Moon and Earth are respectively 2.0×10^{30} kg, 6.0×10^{24} kg and 7.4×10^{22} kg. The distances of the Sun and the Moon from the Earth are respectively 1.50×10^{11} m and 3.8×10^8 m. Calculate the magnitude of the gravitational force exerted on the Earth by the Sun and by the Moon and comment on the ratio of these forces.

$$F_{sun} = (6.67 \times 10^{-11} \text{ N m}^2 \text{ kg}^{-2})\frac{(2.0 \times 10^{30} \text{ kg})(6.0 \times 10^{24} \text{ kg})}{(1.50 \times 10^{11} \text{ m})^2}$$

$$= 3.6 \times 10^{22} \text{ N} \quad \text{(to 2 s.f.)}$$

$$F_{moon} = (6.67 \times 10^{-11} \text{ N m}^2 \text{ kg}^{-2})\frac{(7.4 \times 10^{22} \text{ kg})(6.0 \times 10^{24} \text{ kg})}{(3.8 \times 10^8 \text{ m})^2}$$

$$= 2.1 \times 10^{20} \text{ N} \quad \text{(to 2 s.f.)}$$

Comment: The gravitational field at the Earth from the Sun is about 170 times as strong as the gravitational field from the Moon. This is as we should expect except that the tides are controlled by the Moon much more so than by the Sun and this seems a little odd. The reason is that tides are a response to the difference between the gravitational fields on opposite sides of the Earth. And the Moon's gravitational field changes proportionately more across the Earth's diameter than does the Sun's. This is because the Earth's diameter is a much larger fraction of the Moon's distance than of the Sun's distance. So the Moon controls the tides.

(See also *gravitational constant G, gravitational field strength 'g', gravitational potential (radial fields), planetary motion, satellites, escape speed*.)

Newton's laws of motion are the fundamental principles of Newtonian mechanics. Newton needed to define the relationship between force and motion before he could use the law of universal gravitation to explain the motion of the planets around the sun.

You may be asked to describe an experiment to illustrate one of Newton's laws of motion. The wording sounds a little tentative but it has to be. Newton's laws of motion cannot be verified by experiment. The first law is the worst; it is just not possible to create force-free conditions across a large enough region to be interesting. The failure of any experiment to check the second or third laws will always be attributed to friction and the magnitude of the frictional forces is always calculated by assuming that these forces just account for the disparity between experiment and theory. Fundamental ideas never can be checked but they can be illustrated. Newton's laws are justified by the success of the classical mechanics they support.

An experiment which illustrates one of the laws need do no more than provide some data which more or less fits the mathematical statement of the law. The laboratory experience is part of learning what Newton's laws mean.

The three laws are discussed under *Newton's first law of motion, Newton's second law of motion* and *Newton's third law of motion*.

Newton's second law of motion

'The magnitude of the resultant force acting on a body is proportional to the rate of change of the body's momentum and the direction of the force is the direction of the momentum change.'

Assume that the mass of the body in question is constant. Then, using calculus notation:

$$\text{resultant force} \propto \frac{d}{dt}(mv) \propto m\frac{dv}{dt} = \text{constant} \times m \times a$$

where m is mass and a is acceleration. The unit of force is defined by setting the constant equal to unity:

$$\text{unit of force} = \text{unit for } (m \times a) = \text{kg m s}^{-2}$$

Instead of writing kg m s^{-2}, we write N. We call this unit the newton (N).

The word equation definitions of force and the newton are as follows:

force = mass × acceleration

newton = kilogram metre per second squared (kg m s^{-2})

An account of an experiment to illustrate Newton's second law must include a diagram, a list of apparatus, a list of the measurements (to include how acceleration is found) and how the data is used to establish the conclusion.

Experiment: to illustrate the proportionality of acceleration to applied force for a body of fixed mass

Apparatus

Set up the linear air track arrangement shown in the diagram. The falling mass M consists of several known masses which can be removed and attached to the trolley. This ensures that the total mass of the accelerating system, $M + m$, is constant. The card which is carried by the trolley and which breaks the light beams is 10.0 cm long. The computer can record the time intervals Δt_A and Δt_B for which the light beams at A and B are broken by the card. The time interval t between the arrival times of the card at A and B is also recorded.

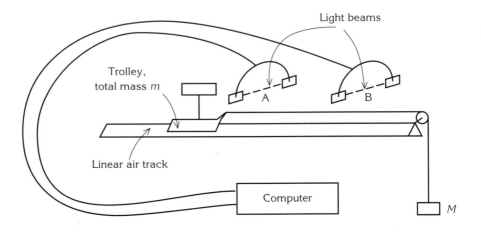

Measurements

For each run of the trolley, the computer records its velocity at A [$v_A = (0.100 \text{ m})/\Delta t_A$], its velocity at B [$v_B = (0.100 \text{ m})/\Delta t_B$] and the acceleration [$(v_B - v_A)/t$]. The value of the acceleration a for each value of the mass M is recorded. Remember to keep the total mass $(m + M)$ constant.

Conclusion

Draw a graph of the applied force Mg against the acceleration a for constant total mass $(m + M)$. If it is a straight line passing through the origin, the proportionality of force and acceleration for constant mass will have been illustrated.

Comment

1. Although the acceleration of the trolley can be measured reliably, it is not possible to increase the applied force by a known fraction without making use of Newton's second law – the very law we are trying to illustrate. An experiment of this sort can be no more than a learning exercise.

2   You must state precisely what data is recorded by the computer and what calculations it performs to determine the acceleration of the trolley. Otherwise the examiner will think that you are using the computer as a cover-up and give you no marks.

3   There are other ways of applying a variable force, such as using a variable number of elastic strings, all stretched by the same amount.

4   A neat alternative method is to attach a trolley to a ticker tape and pull the tape through the vibrator with a constant force using a newton meter or elastic threads. Since the distance moved s in time t is given by $s = \frac{1}{2}at^2$, and since s and t can be found from the tape, the acceleration a for a given force can be calculated. Repeat the experiment for a range of forces and a range of masses.

(See *newton, impulse, experiments.*)

Newton's third law of motion

'Whenever a force acts on a body an equal but oppositely directed force of the same kind acts on a different body.'

Note the following points:

* forces occur in pairs
* the two forces are of the same kind
* they are equal in magnitude
* they act along the same line
* they act over the same time intervals
* they act in opposite directions
* they act on different bodies.

Some of these points are similarities, some are differences.

It can be shown that the law of *conservation of linear momentum* is implicit in Newton's third law of motion. If, when two bodies collide, the forces the two bodies each exert on the other are equal and opposite at each and every instant, then the impulses must be equal and opposite at each and every instant. This means that the momentum gained by one body must just equal the momentum lost by the other. The linear momentum of the two bodies treated as an isolated system is unchanged; it is conserved.

The law of conservation of linear momentum is an absolute (always true). It applies to any collision within a system of colliding bodies in the absence of external forces on the system. It is a consequence not of experiment but of principle. In contrast with this, the law of conservation of mechanical energy may or may not apply in a particular collision.

To illustrate the law of conservation of linear momentum, you would set up some sort of friction-free collision and measure the total momentum in one direction before and after the collision. (Reflection at a wall or from one end of a linear air track will NOT do. Reflection at a wall involves transferring momentum to the Earth and you cannot measure the gain in momentum by the Earth during the collision.)

Experiment: to illustrate the law of conservation of linear momentum

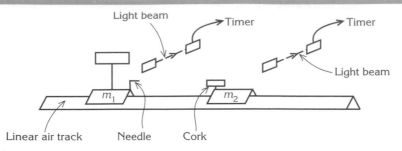

Apparatus

The apparatus is shown in the diagram. Place both trolleys on a linear air track or friction compensated runway. (Always label the runway in a diagram as friction compensated when this is the case.) One trolley carries a 10.0 cm long card mounted at the same height as the two laser beams. Light detectors, which are operated by the laser beams, are attached to digital timers. The timers record the time intervals during which the laser beams are cut by the card on the left-hand trolley. The needle and cork ensure that the trolleys stick together after the collision.

Measurements

Measure the masses m_1 and m_2 of two trolleys by weighing them on a digital balance. Use a ruler to check the length of the 10.0 cm card. Give the trolley with the card a push so that it approaches the second, still stationary, trolley at constant speed. It will collide with the second trolley, stick to it because of the pin and cork arrangement and then travel on with reduced constant speed. Record the values of the time intervals t_1 and t_2 measured by the digital timers. Measure several different pairs of values for t_1 and t_2. Make sure that the collision occurs in between the two time measurements.

Conclusion

The approach speed of the single trolley will be $[0.100/t_1]$ m s^{-1} where t_1 is the time recorded by the first digital timer. The speed of the two trolleys stuck together is $[0.100/t_2]$ m s^{-1} where t_2 is the time recorded by the second digital timer.

According to the law of conservation of linear momentum,

$$m_1 \times (0.100/t_1) = (m_1 + m_2)(0.100/t_2)$$

Check that this equation balances for each pair of time values. Each successful check is an illustration of the law.

(See *impulse*.)

node: a point on a *stationary wave* where the amplitude is zero. It is a point where the two component waves arrive out of phase at all times.

noise might be a sound which is too loud for comfort or without any relevance to the listener. Alternatively, it might be an unwanted (voltage) signal superimposed on a clean signal during transmission, detection or amplification and thereby obscuring the value of the clean signal.

nuclear atom: the model of the atom introduced by Rutherford (1911) in his analysis of the *alpha particle scattering* experiment. The model proposes that the mass content of an atom is almost wholly confined to a central positively-charged nucleus whose diameter is about 10^{-14} m. Atomic radius it typically 10^{-10} m.

The nucleus itself consists of Z protons and N neutrons. The integer Z is called the proton number or atomic number while the integer N is called the neutron number and their sum A (= Z + N) is called the mass number. Protons and neutrons are both *baryons* and they are collectively called nucleons in their role as nuclear particles. Nuclei with one value for Z but different values for N are called *isotopes*.

The alpha particle scattering experiment established the idea of the nucleus but said nothing about the extranuclear electrons. This key development came a year or two later with the publication of Bohr's theory of the hydrogen atom. Bohr assumed that the electrons were held in their orbits by a centripetal force provided by the coulomb attraction between the electron and the nucleus. He assumed further that the angular momentum of the electron (that is, its linear momentum mv multiplied by the orbital radius r) had to equal $nh/2\pi$ where n is an integer and h is the Planck constant. This meant that the electron could occupy only those orbits in which its total energy (potential plus kinetic) satisfied the relation $E_n = -R_{\infty hc}/n^2$ where R_∞ is the Rydberg constant. The minus sign indicates that the state of zero total energy is when the electron is at an infinite distance from the nucleus, with $n = \infty$. Another, later, way of defining these special orbits is to associate the electrons with a wavelength and to argue that the electron orbits must be a whole number of half wavelengths (*stationary wave*).

The normal state of the atom is with the electron in its ground state ($n = 1$). If the atom absorbs just enough energy in a collision to move the electron into a higher, excited state with a higher value for n, then the atom can return to the ground state by emitting a photon of energy $[E_n - E_1]$ equal to $R_{\infty hc}[1/1 - 1/n^2]$. All the possible transitions and the frequencies f of the spectral lines emitted by hydrogen are summarised in the formula:

$$hf = \Delta E = E_n - E_m = R_{\infty hc}\left[\frac{1}{m^2} - \frac{1}{n^2}\right]$$

If we put $m = 1$ and $n = \infty$, we find that the energy needed to remove the electron far away from the nucleus works out as $R_{hc}[1 - 0]$ or, R_{hc}.

$$R_{\infty hc} = (1.097 \times 10^7 \text{ m})(6.63 \times 10^{-34} \text{ J s})(3.00 \times 10^8 \text{ m s}^{-1}) = 13.6 \text{ eV}$$

This is the energy needed to ionise the hydrogen atom. Hence 13.6 V is called the ionisation potential.

The discrete energy levels occupied by the extranuclear electrons lead to the concepts of *excitation* and *ionisation* and to the distinction between *emission spectra* and *absorption spectra*. See *N–Z curve for stable nuclei*.

nuclear matter is the collection of neutrons and protons that makes up a nucleus. The positive charges on the protons should ensure that nuclear matter will blow itself apart. But this is to ignore the influence of the strong interaction, which is about a hundred times stronger than the electromagnetic interaction over the distances involved in the nucleus. The weak interaction acts within nucleons, not outside them. (See *four fundamental interactions*.)

nuclear power is invariably provided by controlling a chain reaction consisting of numerous identical neutron-induced *fission* reactions, such that each fission reaction triggers another identical reaction in sequence. No reliable way has yet been found of employing a *fusion* process to supply power in a controlled manner.

The related mathematics involves the use of three non-SI units: the *unified atomic mass constant*, the energy unit MeV (see entry) and a related mass unit MeV/c^2 (see entry). The energy released is calculated by considering changes in the *mass defect* and *binding energy* together with the *mass–energy equivalence* concept.

The technology of nuclear power is built around the *fuel rod*, the *moderator* and the *coolant* collected together in the *thermal fission reactor*.

nuclear radius can be calculated from the formula:

$$R = r_0 A^{1/3}$$

where A is the mass number and r_0 is approximately equal to 1.2×10^{-15} m. This result is confirmed within limits by electron and proton *scattering* experiments; it is not exact. It implies that nuclear density, at around 10^{17} kg m^{-3}, is much the same for all atoms.

nuclear reactions: these occur when the *nucleus* of an atom is struck by a fast nuclear particle or a high energy photon which is then absorbed. The fast particle or photon, together with the energy it carried, usually renders the transformed compound nucleus unstable. If stability is reached by the emission of a particle which is different from the incident particle then a *nuclear transformation* will have taken place.

Nuclear reactions are often written out in the following conventional way:

initial nuclide (incident particle, emergent particle) final nuclide

The following nuclear reactions are all of some historical significance.

The first artificial transmutation (Rutherford, 1919) was achieved by bombarding nitrogen with alpha radiation, thus forming some oxygen which was detected spectroscopically:

$${}^{14}_{7}N({}^{4}_{2}\alpha, {}^{1}_{1}p){}^{17}_{8}O$$

The first transmutation triggered by a particle from an accelerator (Cockcroft and Walton, 1932) was the changing of lithium into helium by bombarding it with protons:

$${}^{7}_{3}Li({}^{1}_{1}p, {}^{4}_{2}\alpha){}^{4}_{2}He$$

The neutron was discovered (Chadwick, 1932) by bombarding beryllium with alpha particles and transforming it into carbon:

$${}^{9}_{4}Be({}^{4}_{2}\alpha, {}^{1}_{0}n){}^{12}_{6}C$$

nuclear reactor: see *thermal fission reactor*.

nuclear transformation: a change in the structure of a nucleus which may be brought about by the spontaneous emission of a nuclear particle (*radioactive decay*) or by bombarding a nucleus with fast nuclear particles which are absorbed and so bring about a change of nuclear structure (*nuclear reaction*).

nucleon: either a proton or a neutron.

nucleus: the dense, central portion of an atom. It is made up from neutrons and protons. It has a density around 10^{12} times the average density of the atom and a diameter about ten thousand times smaller than the atomic diameter.

nuclide: a *nucleus* for which *mass number* and *atomic number* are specified.

Nyquist's theorem states that the sampling frequency needed to monitor a signal component of frequency f_m must be greater than $2f_m$. This is the theoretical minimum below which the required information cannot be recovered. In practice, the sampling frequency needs to be around $10f_m$ or greater.

N-Z curve for stable nuclei is a chart, not a graph, of *neutron number N* against *proton number Z*. Each square on the chart represents a particular N–Z pair. The diagram is called a chart because the allowed N, Z values are integers only and it makes no sense to connect them together with lines. Most squares represent nuclei which are too unstable ever to be produced; this is why most of the squares are empty. The three thousand or so known nuclei occupy squares in the curved fan-shaped strip. It is curved because the N values rise slightly faster than the Z values. Only about 10% of the known nuclei are found in nature. The rest are produced artificially in nuclear reactors or particle accelerators. The largest stable nucleus belongs to bismuth, Z = 83, N = 126. Z for artificially produced nuclei can exceed 100.

The following statements are broadly correct but there are occasional exceptions. Stable nuclei are found mostly along the centre of the strip. Nuclei above this strip are unstable and are brought back towards the centre through radioactive decay. This will be by beta-negative decay if they are of low Z number (one neutron less and one proton more) or by a mixture of alpha decay and beta-negative decay for high Z number. Unstable nuclei below the stable strip are subject to K-capture (mostly) and to beta-positive decay (less often). Beta-positive decay was a long time being discovered because most of the nuclei that decay by this means are artificial. K-capture describes the combination of an inner K-shell electron with a proton in the nucleus to form a neutron. The electron shells have to re-adjust and the right number of electrons for this to happen are still present. The three nuclear decay process are shown on the chart. See also *alpha decay*, *beta decay*.

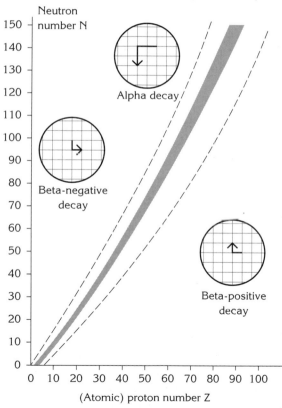

N–Z *curve for stable nuclei*

object position: the source of light in a *ray diagram*.

ohm: the SI unit of *resistance*. The symbol used is Ω. The ohm is defined by the word equations:

$$\text{resistance} = \frac{\text{voltage}}{\text{current}}$$

$$\text{ohm} = \frac{\text{volt}}{\text{ampere}}$$

ohmic conductors are ones which obey *Ohm's law*. They are sometimes called linear components. Non-linear components are sometimes described as 'active'; the discriminating behaviour of many electronic circuits usually begins with an active or non-ohmic component. Control and stability are achieved partly by feedback and partly by using non-ohmic conductors.

Ohm's law states that, for a conductor held at constant temperature, the current I flowing in it is proportional to the potential difference V across its ends. The ratio of potential difference to current is called the resistance R of the conductor. Ohm's law is generally written in the form:

potential difference in volts = current in amps × resistance in ohms

$$V = IR$$

Ohm's law can be written down for a complete circuit. In these circumstances the potential difference becomes the circuit e.m.f. E and Ohm's law is written:

$$E = IR$$

where R is the equivalent resistance of the circuit as a whole. (The logic gets a bit involved. We would really be saying that the e.m.f. is the sum of the voltage drops round the circuit and that Ohm's law enables us to say of each voltage drop that $V = IR$, whence ...)

Ohm's law is an odd statement in that it applies to those conductors to which it applies. This does not seem very helpful. However, Ohm's law applies to a large enough set of components for it to be an important starting point in any analysis. The mathematical statement of the law (the equation) has no constant in it. This is because the equation which states Ohm's law is effectively a definition of the ohm.

oil drop experiment: see *Millikan's oil drop experiment*.

oil film experiment: an experiment in which a small drop of oil, of known diameter, is placed carefully on the surface of clean water in a tray. The water in the tray has been previously dusted with fine powder so that the area over which the oil drop spreads can be measured. The oil drop is about half a millimetre in diameter and this can be measured with a millimetre scale and a good magnifying glass. If d is the diameter of the oil drop and if A

is the area of the oil film on the water, the diameter of the oil molecule cannot be larger than t where

$$t = \left[\frac{4}{3}\pi\left\{\frac{d}{2}\right\}^3\right] \div A$$

This method of measuring 'molecular diameter' is no more accurate than an order of magnitude for molecular size. It is not even much good for this because oil molecules are long and thin and end up packed vertically. But it is simple, direct and tells us what sort of magnitudes are involved.

Olber's paradox states that if the universe is infinite and uniformly filled with stars, then a star will be seen in any direction we care look and the whole sky will be as bright as the surface of an average star.

In reality, in a universe of finite age of 10^{10} years say, only those stars within a radius of 10^{10} light years will be visible. The comparatively low density of stars makes the light level tolerable. This resolution of the paradox is consistent with a big-bang theory but not with a steady state theory of the universe.

optical fibre systems are signalling systems which rely, at some point, on the transmission of light pulses along lengths of narrow glass fibres. The fibres are mechanically faultless (otherwise they would already be broken), flexible (because they are so thin) and almost 100% transparent.

Optical fibres are about an eighth of a millimetre across and consist of core and cladding. Both core and cladding are made of glass and in older fibres the core diameter is a little more than half the total diameter. The cladding has a lower refractive index than the core: this ensures *total internal reflection* of oblique light rays. The change in refractive index can be either stepped (i.e. sudden) or graded (i.e. smoothed out).

A sharp pulse entering one end may arrive at the other end spread out in time. There are two reasons for this time dispersion. First, the light can travel over a small range of zigzag paths at slightly different angles; the steeper the angle, the further the light has to travel. Secondly, the longer wavelength components travel faster than the shorter wavelength components. Time dispersal places a serious limit on the information carrying capacity of the fibre.

Later fibres, still with a total diameter of about an eighth of a millimetre, have a core diameter of only 5 µm. Since there is now no chance of any zigzagging, the fibres are called monomode fibres. Also, if the light comes from small laser chips whose light output is essentially monochromatic, both sources of time dispersion are much reduced and the information carrying capacity is much improved.

ordered and disordered energy are the two forms of incoming energy referred to in *thermodynamics (first law of)*. Ordered energy is like gravitational potential energy, kinetic energy of a single moving mass or energy stored in a stretched spring. It can be transferred from one system to another wholly in the form of useful work. Disordered energy is energy stored in the form of random molecular motion. No more than a fraction of this can be transferred in the form of useful work; the remainder is transferred as disordered energy in a heating action.

A B C D E F G H I J K L M N O P Q R S T U V W X Y Z

173

packing describes the three-dimensional pattern taken up by atoms within a crystal. Face-centred cubic and body-centred cubic structures are examples. Bonds may be either *ionic* or *covalent*. Within close packing the density of the crystal moves towards its maximum value, as in the hexagonal close packed structure of materials such as zinc. The tetragonal structure of diamond is a rare example of high crystal strength coupled with an open packing arrangement. The high strength is linked to the purity of the carbon.

pair production relates to the formation of a *particle–antiparticle* pair from the energy provided by a single gamma-ray *photon*. The reaction takes place within the vicinity of a particle such as a nucleus, whose recoil helps to conserve energy and momentum in the transformation event. The opposite process, called annihilation, occurs at the meeting of a particle with an antiparticle. The energy and momentum are carried off by two gamma-ray photons.

(Pair production has nothing to do with the formation of electron–hole pairs, which appear in semiconducting materials when an electron is freed from a fixed atom in the lattice by an excitation process and leaves behind a hole. The hole is no more than a vacancy which is mobile in an intrinsic *semiconductor* but static if it is linked to an impurity atom in an extrinsic semiconductor.)

parallax error is the random error associated with identifying the position of a point when observing it obliquely (e.g. the reading for a needle above a scale). It is a random error and the average of several independent measurements should be close to the correct value. If a parallax error is a consequence of a poor arrangement of the apparatus, it becomes a systematic error. Parallax errors should be minimised wherever possible by using mirrors. See *random error*.

parallax (stellar) is the angle θ shown in the diagram. That is, the parallax angle for a star is the maximum angle between the Earth and the Sun as seen from the star. Since the radius of the Earth's orbit round the Sun is known accurately, the distance of the star can be calculated, once θ is known.

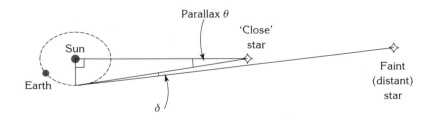

The angle of parallax is found by measuring the angle δ between the star and a nearby faint star over a period of several months. The actual calculation of the angle of parallax from changes in the value of the angle δ is complicated, mainly because neither the bright star nor the faint star will be in the plane of the Earth's orbit about the Sun. The calculation is a three-dimensional affair. Measurements for one bright star against several faint stars are taken and averaged. The method assumes that the bright stars are the stars nearest the Earth and that faint stars are much more distant. Should the bright star also be a distant star, the angle δ will not vary as the months go by.

Useful parallax measurements can be made for no more than about the two thousand or so stars nearest to the solar system, that is, for one star per hundred million stars in this galaxy. And there are about 10^{13} galaxies. These relatively few measurements allow the absolute brightnesses of these stars to be determined from their apparent brightnesses. Absolute brightness can then be correlated with surface temperature using *Wien's law*. This is the basis for the *Herzsprung–Russell diagram*. All other distance measurements used by astronomers hinge one way or another on these first parallax measurements.

parallelogram law: the law applying to the *addition of vectors*. Each of the vectors is represented in magnitude and in direction by the sides of a parallelogram. The included diagonal of the parallelogram represents, in magnitude and in direction, the resultant of the two vectors. The other diagonal equals the difference between the two vectors.

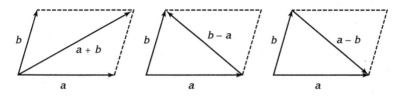

parallel plate capacitor: a device for storing electric charge constructed from two parallel conducting plates mounted a short distance apart.

The diagram below shows a parallel plate capacitor with the distance between the plates exaggerated so that the *field lines (electric)* can be shown clearly.

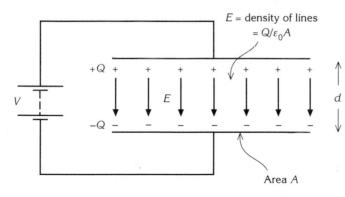

The *capacitance* of the arrangement is defined by the equation:

$$\text{capacitance} = \frac{\text{charge stored on one plate}}{\text{potential difference between plates}}$$

All the lines of force go from the positive charge +Q on the upper plate directly to the negative charge –Q on the lower plate. The number of lines of force is Q/ε_0 since there are $1/\varepsilon_0$ lines of electric force per unit charge. The *electric field strength* E is the number of lines of force per unit area. If the area of the plates is A, then:

$$E = \frac{Q}{A\varepsilon_0}$$

But

$$E = \frac{V}{d} = \frac{Q}{A\varepsilon_0}$$

and since $C = Q/V$

$$C = \frac{\varepsilon_0 A}{d}$$

If the space between the plates is filled with insulating material, the capacitance of the arrangement changes by a numerical factor ε_r where ε_r is called the *relative permittivity* of the insulating substance. The equation for the capacitance becomes:

$$C = \frac{\varepsilon_r \varepsilon_0 A}{d}$$

How this increase in the capacitance comes about is indicated in the diagram below.

On the left is a capacitor filled with vacuum. Charge Q is stored and the field between the plates is E. On the right is an identical capacitor filled with an insulating material. The electric field polarises the insulator. Many of the lines of force which set out from the positive plate end on the polarised charge on the dielectric. A small positive charge carried from B to A would have to fight against a reduced number of lines of force for most of the way. The charge on the plates is the same but the electric field strength and the potential difference between the plates are both reduced. Since C = Q/V, the capacitance must increase. The factor by which it increases is called the relative permittivity of the insulator.

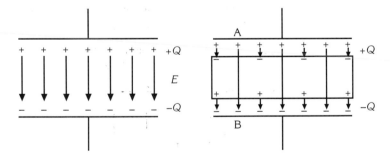

paramagnetic substances are weakly magnetised by a surrounding magnetic field, in the same direction as the field. The *relative permeability* of a paramagnetic substance is slightly greater than unity. A bar of paramagnetic substance will align its long side parallel to a magnetic field if freely suspended. *Ferromagnetic* materials become paramagnetic above the *Curie temperature*.

particle accelerators provide beams of high energy particles, especially electrons and protons, for scattering and collision experiments. They are usefully divided into linear accelerators and ring accelerators.

A B C D E F G H I J K L M N O P Q R S T U V W X Y Z

Electron gun. This consists of a heated filament which provides electrons by *thermionic emission*, a negative cathode to give them direction and two cylindrical anodes at different voltages to focus and accelerate the beam.

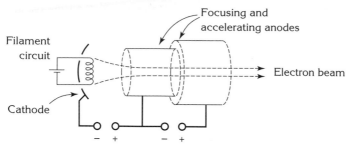

Linear accelerators. These machines are based on the principle that the kinetic energy of a negatively-charged particle will increase when it passes from a region of low electrical potential into a region of higher electrical potential. For protons, the voltages will be the other way round.

High speed electrons from a Van der Graaff generator (with an electron gun within the sphere) are projected through a series of conducting cylinders. The speed of the electron increases as it leaves one cylinder and enters the next provided there is an increase in potential. This is achieved without recourse to enormous voltages by changing the directions of the voltages while a bundle of electrons is within a cylinder. With the changes happening at equal intervals of time, the lengths of the cylinders are progressively increased to accommodate the increasing electron speeds. The final cylinder length is measured in kilometres. The cylinder lengths have to take the relativistic speeds of the electrons into account.

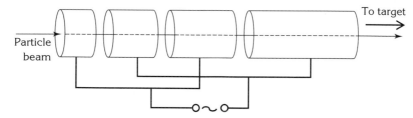

Ring accelerators. A proton moving with speed v at right angles to a magnetic field of strength B follows a circular path of radius r, and $r = mv/Be$ (see *force on moving charge*). This principle is adopted in the cyclotron to accelerate protons or ions to high energies.

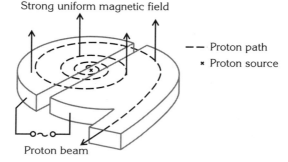

The cyclotron is similar to a linear accelerator in that the proton speed is boosted each time the bunch of protons passes from one D section to the other and that the direction of the voltage between the two D sections changes with a frequency which matches the protons' arrival times. The changing proton speed is accommodated by the increase in the length of its semicircular path. The frequency f of the AC power source which controls the direction of the potential difference between the D sections is as follows:

$$f = \frac{v}{2\pi r} = \frac{Be}{2\pi m}$$

Electrons reach relativistic speeds at too low an energy for the cyclotron to be useful. Synchrocyclotrons are used instead.

particle interactions is a phrase that refers to what happens when two or more *fundamental particles* collide, mostly at high energies in a *particle accelerator*. In this context, particle decay is treated as an interaction.

The outcome of a collision must satisfy a number of *conservation laws* and these laws must take account of the creation of new particles and the annihilation of incident particles. Apart from conservation of mass–energy, linear momentum and angular momentum, there are nuclear conservation laws characterised by quantum numbers. These include charge Q, baryon number B and strangeness S. Conservation of charge is nicely illustrated by electron–positron annihilation or by pair production. In both cases the total charge before and after the event adds up to zero. Conservation of baryon number implies that, when second and third generation baryons decay, they must decay (eventually) into protons or neutrons, thereby conserving baryon number. Mesons consist of quark–antiquark pairs and have baryon number zero. There is therefore no barrier in respect of baryon number to the creation and annihilation of mesons – but charge must be conserved, so either the mesons are neutral or they come and go in positive–negative pairs. Strangeness S is more trouble. It is conserved in events controlled by the strong interaction but not in events controlled by the weak interaction.

Lepton number is another conserved quantity. Remember that lepton number L, like baryon number B, is a quantum number which may have value +1, –1 or zero. It is not the number of leptons. In the beta decay of neutrons, for example, the lepton numbers of the electron and the antineutrino on the output side are +1 and –1, respectively. The electron and proton have Q values of –1 and +1; the proton and neutron both have baryon number B equal to 1. See *baryons*, *hadrons*, *leptons*, *mesons* and *proton decay and neutron decay*.

particle physics is the study of *subatomic particles*, including *fundamental particles*. We live in a low-energy environment and the subatomic particles that mostly matter are the *nucleus*, the *proton*, the *neutron*, the *electron* and the *photon*. Atoms, molecules and solid materials need a low-energy environment, such as we find on the Earth's surface, for their very existence. The inside of a star is a high-energy environment in which atoms cannot exist. Far higher energy environments are needed for the generation of the more massive fundamental particles; it is these environments that are mimicked by the collisions inside *particle accelerators*.

Protons and neutrons are members of a large group of compound particles known as *hadrons*. Hadrons, which consist either of two quarks (*mesons*) or of three quarks (*baryons*), are subject to all four of the four fundamental interactions. The strong interaction acts only on quarks and, for this reason, the strong interaction acts on hadrons but not on *leptons* or *exchange particles*.

Leptons, such as the electron or positron, are immune to the strong interaction but can and do react with hadrons through the weak interaction.

The photon is an example of a third group of fundamental particles known as *exchange particles*. (See *conservation laws (nuclear)*, *Feynman diagrams*, *particle interactions* and *radiation detectors*.)

path difference is the extra distance one wave motion travels further than another; the two distances are measured from points where the two waves are in phase and end at the point where the two waves meet. The *phase difference* at the meeting point is calculated as follows:

$$\text{phase difference} = \frac{2\pi}{\lambda} \times \text{path difference}$$

The wavelength value can be omitted if the path difference, as is often the case, is expressed as a multiple of the wavelength.

The diagram below shows an example in exaggerated form.

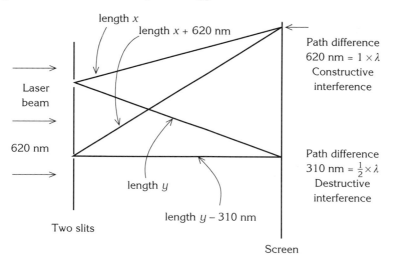

length x

length $x + 620$ nm

Laser beam

620 nm

length y

length $y - 310$ nm

Two slits

Path difference
620 nm = $1 \times \lambda$
Constructive
interference

Path difference
310 nm = $\frac{1}{2} \times \lambda$
Destructive
interference

Screen

Path difference is usually the first thing to be calculated when determining or explaining the *interference* pattern formed where two waves meet. (See also *coherence*.)

penetrating power: see *attenuation*.

period is the time taken up by one complete oscillation of a vibrating system. More formally, it is the time interval between successive occasions when a body whose motion is oscillatory moves through the same point in the same direction. The SI unit of period is the second (s), the symbol used, T.

It is general practice to use period when discussing simple mechanical oscillations but to use frequency when dealing with complex mechanical or electrical oscillations. This is because such systems are driven by frequency generators.

permeability of free space, μ_0, is a constant which appears in any expression which relates the magnetic field strength B at a point to the current I flowing in a nearby conductor. Examples are $\mu_0 I / 2\pi r$ near a straight wire and $\mu_0 n I$ within a solenoid.

The SI unit for μ_0 is henry per metre (H m^{-1}). The magnitude of μ_0 is defined as $4\pi \times 10^{-7}$ H m^{-1} as a step towards the definition of the *ampere*. (See *speed of light*.)

permittivity of free space is the constant ε_0 which appears in *Coulomb's law*. The magnitude of ε_0 is set by defining the magnitudes of μ_0, the permeability of free space, and of *c*, the speed of light. These definitions put the value of ε_0 at $8.854\ 187\ 817 \times 10^{-12}$ F m^{-1}. This comes about because electromagnetic theory shows that:

$$c = \frac{1}{\sqrt{(\mu_0 \varepsilon_0)}}$$

phase is an angle in *radians*, between 0 and 2π, which indicates a fraction of a cycle completed. (See *phase difference*.)

phase difference is an angle in radians, between 0 and 2π, which indicates the extent to which two wave motions of the same frequency are out of step. It is the fraction of a complete cycle, in radians, by which one signal would have to be retarded or advanced to be in phase with another signal of the same frequency.

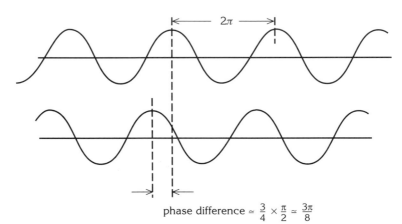

phase difference $\approx \frac{3}{4} \times \frac{\pi}{2} \approx \frac{3\pi}{8}$

The definition and its meaning are illustrated in the diagram above. The lower signal leads the upper signal by about $3\pi/8$ radians. 'Leading' seems the wrong way round until we remember that time later, is on the right. The upper signal peaks later; it lags the lower signal. The concept is particularly relevant to alternating currents and voltages.

(See *path difference, interference, coherence, wavefront*.)

photoelectric effect: the enforced expulsion of *electrons* from a metal surface when the surface is irradiated with *electromagnetic radiation* of sufficiently high frequency. For most metals, this frequency lies in the ultraviolet. The alkali metals show the effect with visible radiation, which for sodium is in the range we see as green. There are semiconducting materials and metals with special coatings which release photoelectrons for the full span of the visible spectrum and into the near infra-red.

A beam of electromagnetic radiation consists of innumerable *photons*. If the beam is monochromatic with frequency *f*, all the photons in the beam have energy *hf* where *h* is the *Planck constant*, 6.626×10^{-34} J s. If the radiation is incident upon a metal surface, the occasional photon will release all its energy to a single conduction electron. Some of these conduction electrons will escape from the metal and are then called photoelectrons.

In normal circumstances, conduction electrons are free to roam within the metal but they cannot escape. To get out they must pay a sort of 'energy exit fee' called the work function . The work function (symbol ϕ) differs from metal to metal but is a constant for a particular metal.

If v is the maximum speed acquired by a photoelectron and if m_e is electron mass, then a simple application of the conservation of energy law gives:

$$hf = \phi + \tfrac{1}{2}m_e v^2$$

This is known as Einstein's photoelectric equation. It was published in 1905 and verified experimentally in 1912, following the publication of an accurate value for m_e calculated from a value for e produced by *Millikan's oil drop experiment* and earlier measurements of e/m_e. Millikan himself, reported a detailed experimental investigation of Einstein's equation using the alkali metals in 1916. The principal features of his apparatus are shown in the diagram below:

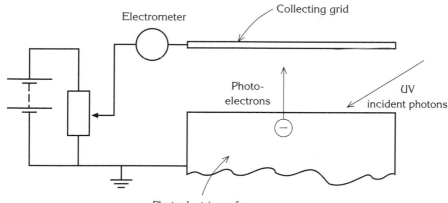

Using monochromatic radiation of frequency f, Millikan measured the minimum negative voltage between the collecting electrode and the earthed photoelectric surface for the photoelectric current to become zero. This is called the stopping potential V_s. Einstein's equation can be re-written:

$$hf = \phi + e.V_s$$

or

$$V_s = \frac{h}{e}f - \frac{\phi}{e}$$

Results obtained for barium were as follows:

Stopping potential in volts	Wavelength in mm	Frequency in 10^{15} Hz
2.40	0.254	1.18
1.47	0.313	0.958
0.91	0.365	0.821
0.57	0.405	0.740

From which the graph on page 182 has been drawn:

The gradient of the line is 4.16×10^{-15} volt seconds. It equals h/e. The value for Planck constant comes out at $(4.16 \times 10^{-15}$ V s$)(1.60 \times 10^{-19}$ C$)$ or 6.67×10^{-34} J s (to 3 s.f.). The negative intercept is 2.51 V. This equals ϕ/e. The work function ϕ for barium comes out at 4.0×10^{-19} J (to 2 s.f.) or 2.5 eV (to 2 s.f.).

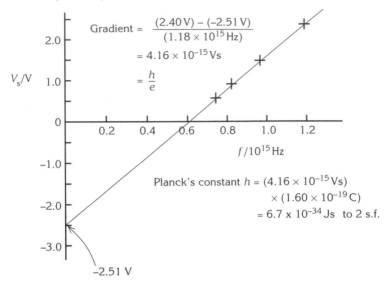

$$\text{Gradient} = \frac{(2.40\,\text{V}) - (-2.51\,\text{V})}{(1.18 \times 10^{15}\,\text{Hz})}$$
$$= 4.16 \times 10^{-15}\,\text{Vs}$$
$$= \frac{h}{e}$$

Planck's constant $h = (4.16 \times 10^{-15}\,\text{Vs})$
$\times (1.60 \times 10^{-19}\,\text{C})$
$= 6.7 \times 10^{-34}\,\text{Js}$ to 2 s.f.

The cut-off frequency equals the work function ϕ divided by the Planck constant h and comes out at 0.60×10^{15} Hz.

The first graph below illustrates the way photoelectric current varies with the voltage on the collector grid, for different values of illumination. The second graph underlines the proportionality between the photoelectric current and the light intensity for light of fixed frequency above the cut-off frequency and for a fixed collector grid voltage above the stopping potential.

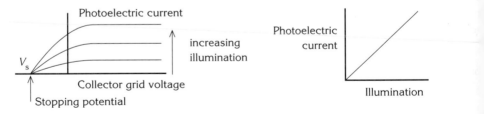

photon: a quantum of electromagnetic radiation. Each quantum of frequency f has energy hf, where h is the *Planck constant*, 6.626×10^{-34} J s. (See also *lepton, wave-particle duality photoelectric effect.*)

photovoltaic cells convert energy from solar radiation directly into electrical energy. The efficiency of such processes is around 10% and the mechanism involves the release of loosely bound electrons in the vicinity of the interface between two semiconducting layers (one p-type, the other n-type). The average energy possessed by the electrons generates a voltage of about half a volt between the two metal electrodes on the far sides of the semiconducting layers. The source of the energy for any current flow is the photon energy employed to release the electrons.

physical quantity: any physically measurable quantity. Temperature, density and electrical potential are physical quantities; appearance, usefulness and worth are not.

pitch is the subjective response of the ears to *sound* frequency. The ears hear frequencies within the range from about 10 Hz to about 15 kHz: around three orders of magnitude. The upper limit comes down with increasing age. Young children can hear up to 20 kHz while adults can only manage 12kHz to 15 kHz. Many older people cannot hear bird song. See *ear*.

planetary motion is the orbital motion of planets about the Sun. The orbits are assumed to be circular. The first step in a problem is almost always to write down that the gravitational force holding the planet in its orbit is the *centripetal force*. If M is the mass of the Sun, if m is the mass of the planet and if r is the radius of the planet's (imaginary) circular orbit, if v is the speed of the planet and if T is its orbital period, then:

$$G\frac{Mm}{r^2} = \frac{mv^2}{r} = \frac{m}{r} \times \left(\frac{2\pi r}{T}\right)^2 = \frac{4\pi^2 mr}{T^2}$$

so $$T^2 = \frac{4\pi^2}{GM} \times r^3$$

or $$T^2 \propto r^3$$

This is Kepler's third law of planetary motion which states that the square of the period of a planet's orbit about the sun is proportional to the cube of its average distance from the sun.

plastic deformation is the continuing flow and change of structure within a material, leading to a drastic increase in strain when a stress close to the breaking value is applied. Where movement within the material centres around countless *dislocations*, the plastic deformation appears to be uniform throughout the specimen. Where deformation is primarily due to slippage between crystal planes, it tends to be more localised and the specimen will be closer to breaking. A material not subject to plastic flow is described as *brittle*. A material in which the onset of plastic flow occurs at high values of stress is described as *strong*. A metallic, crystalline material such as copper, which can endure large plastic deformations, allows dislocations to move freely within its volume and will accommodate appreciable *slip*. It will be both *malleable* and *ductile*. See *stress–strain curve*.

plasticity is the ability of a material to retain any change of shape or size once a deforming stress has been removed. It is the opposite of *elasticity*. (See also *stress–strain curves*.)

Poiseuille's equation states that the volume per second V/t of fluid flowing along a straight pipe of constant radius r and of length l is given by:

$$\frac{V}{t} = \frac{\pi}{8\eta} r^4 \frac{p}{l}$$

where p is the pressure difference between the two ends of the tube and η is the coefficient of viscosity.

The strong dependence of flow rate on tube radius (halve the radius and the flow rate drops to a sixteenth of its former value) underlines the crucial role of vein and artery diameters in animals, xylem and phloem sizes in plants and particle size in soils.

polarisation phenomena are those properties taken on by light and other *electromagnetic waves* when the direction of the transverse wave vector is the same for all parts of the wave. The diagram below shows a rectangular sheet of *Polaroid* cut in half after two parallel reference lines have been marked on it. A beam of ordinary light from a lamp is shown passing through the two sheets of Polaroid. If the two reference lines are parallel, the light gets through both sheets. If the reference lines are at right angles, the light gets through the first sheet but not the second.

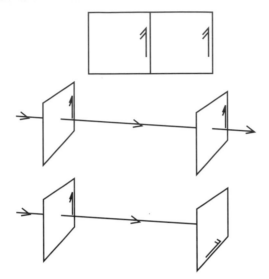

The first sheet of Polaroid is called the polariser, the second sheet is called the analyser and the light between the two sheets is polarised light.

The next diagram shows a glass container filled with a solution of an optically active material such as a solution of dextrose placed in the path of the polarised light. This time, the light should be monochromatic. If the analyser is set for darkness before the container was put in place, then some light will now get through. The light can be absorbed once more by rotating the analyser. Experiment shows that the angle through which the analyser is rotated for darkness is proportional to both the path length in the solution and its concentration. By comparing the rotations with solutions of known and unknown concentrations, it is possible to measure quickly the concentration of any optically active solution.

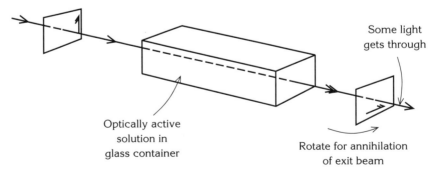

Some light
gets through

Optically active
solution in
glass container

Rotate for annihilation
of exit beam

The same set-up can be used for stress analysis. A perspex model is made of a component the design of which is incomplete, and this is placed between the polariser and the analyser. A large number of dark lines will be seen within the perspex component and these lines will be heavily concentrated in the areas of maximum stress. In this way, the design can be modified by trial and error to ensure stresses are distributed evenly throughout the component.

Polaroid is the *brand name* (hence the capital P) for a nitrocellulose sheet covered with crystals of quinine iodosulphate. All the crystals are lined up in the same direction so that normal light passing through the sheet comes out plane polarised. There is one particular orientation of the sheet at which polarised light passes through unaffected (apart from some modest absorption). Then, starting from this position, as the polaroid is slowly rotated about an axis parallel to the light, it is increasingly absorbed. The peak absorption position corresponds to an angle of rotation of one right angle from the minimum absorption position.

polycrystalline solids are crystalline solids in which the individual crystals are tiny compared with the dimensions of the whole and the *physical properties* of which depend, in large part, on the manner in which the individual crystals are linked together. In metals, the small crystals are known as grains. *Annealing* (heat to a high temperature and allow to cool slowly) encourages a growth in grain size and a softer, more pliable metal.

polymeric solids are formed from long chains of carbon atoms linked together by strong *covalent* bonds. Each carbon atom in the chain is left with two more bonds which hold more atoms along the sides of the string. The diagram below shows two polymeric molecular structures, polyethylene and polyvinyl chloride. The physical characteristics of the solid depend on the nature of the cross-linking bonds which hold the long molecular chains together. (See *thermoplastic polymers*, *thermosetting polymers*.)

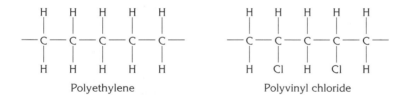

Polyethylene   Polyvinyl chloride

position: an identifiable point on a line or within an area or in a space.

positron: a fundamental particle, identical in mass to the *electron* and bearing an equivalent but opposite (positive) charge.

potential (electric) at a point equals the work done per coulomb bringing a small positive charge from a point at infinite distance from other charges to the point. It is a *scalar quantity* and the symbol is *V*. The SI unit is the volt (V). The volt is another name for the joule per coulomb ($J\ C^{-1}$).

Electric potential is zero at infinite distance and negative near a negative charge but positive near a positive charge. The electric potential *V* a distance *r* from a positive charge *Q* is calculated from the formula:

$$V = \frac{Q}{4\pi\varepsilon_0 r}$$

Electric potential decreases with distance from a positive charge but increases with distance (gets less negative) from a negative charge.

potential (gravitational) at a point equals the work done per kilogram bringing a small mass from a point at infinite distance from other masses to the point. Because gravitational forces are always attractive, gravitational potential is zero at infinite distance and negative elsewhere. It is a *scalar* quantity. The symbol used is *V*. The SI unit is the joule per kilogram (J kg⁻¹). (See *gravitational potential (radial fields)*, *Newton's law of gravitation*.)

potential difference (electrical) between two points is the work done per coulomb of charge transferring a small positive charge from one point to the other. It is a *scalar quantity*. The symbol used is *V*. The SI unit is the volt (V) which is another name for the joule per coulomb (J C⁻¹). Definitions can usefully be stated as the word equations:

$$\text{potential difference} = \frac{\text{work done}}{\text{charge}}$$

$$\text{volt} = \frac{\text{joule}}{\text{coulomb}}$$

potential difference (gravitational) between two points is the work done per kilogram of mass transferring a small mass from one point to the other. It is a *scalar quantity*. The symbol used is *V*. The SI unit is the joule per kilogram (J kg⁻¹).

The Earth's gravitational field, near the Earth and on a scale small compared with the Earth's radius, is uniform. Gravitational potential is increasingly negative downwards. Raising a body of mass *m* a vertical distance *h* moves it to a point of increased gravitational potential. The increase in gravitational potential is *gh*. The gravitational potential difference between the higher point and the lower point is *gh* and the body gains potential energy *mgh* going from the lower point to the higher point.

It is usually easier to work with scalar than vector quantities, since with the latter direction adds another complication. (See *potential (gravitational)*.)

potential divider: a fixed resistor which may be tapped at any point along its length to enable part of the *potential difference* across the whole resistor to be accessed. The circuit below shows how a potential divider may be used to adjust the voltage across a lamp between zero and any pre-decided value.

The voltage across the lamp or any other component will not, in general, vary linearly with the distance moved by the slider. This will happen only when the resistance of the component is much larger than the end-to-end resistance of the potential divider. The series resistor R is a protection device. It may protect the component, the lamp, from the full terminal potential difference of the power supply. It may protect the power supply from the danger of delivering too much current if the component is connected straight across its output. Whatever the reason, its value must be computed beforehand.

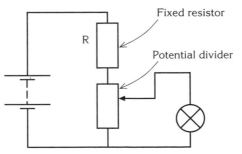

Fixed resistor

R

Potential divider

potential energy is the energy stored in a system by reason of its position or its condition. It is a *scalar quantity*. The SI unit is the joule (J). The symbol used may be the energy symbol W or there may be a subscript W_p to distinguish it from kinetic energy W_k. The forms E_p and E_k are also used. A mass m raised a height h above the ground has potential energy mgh joules. Potential energy can also be stored as elastic strain energy, as electrical energy in a capacitor and as one component of internal (thermal) energy in solids and liquids.

The expression mgh used for gravitational fields is misleadingly simple. Height h is measured from a level at which potential energy is assumed to be zero. This is sufficient for many purposes. Precise work requires us to set the potential energy zero for a mass in a gravitational field at infinite distance from the centre of the field. Because potential energy depends on the mass and on the position in the field, we remove the mass dependence by introducing the concept of potential. Potential energy released or stored equals (mass × change of potential).

This approach works for electric fields too. (See *gravitational potential (uniform fields)*, *gravitational potential (radial fields)*.)

power is rate of energy transfer. It is a *scalar quantity*. The SI unit of power is the watt (W). The symbol for power is P. Power and the watt are defined by the word equations:

$$\text{power} = \frac{\text{energy transferred}}{\text{time}}$$

$$\text{watt} = \frac{\text{joule}}{\text{second}}$$

The topic is made more complicated for the A level student by the variety of ways in which 'energy transferred' has to be calculated. A simple problem in mechanics is solved by multiplying a force by a displacement and dividing by a time (all in SI units!). But displacement divided by time is velocity (usually constant in these circumstances) and power becomes the product of force and velocity. When, for instance, a 200 g mass is falling towards the ground then, at the instant when its speed is 12 m s^{-1}, the gravitational field is working at a rate of:

$$\text{rate of working} = \text{force} \times \text{velocity}$$
$$= (0.200 \text{ kg} \times 9.8 \text{ m s}^{-2}) \times (12 \text{ m s}^{-1})$$
$$= 23.52 \text{ W}$$

At the starting point of the fall, the rate of working of the gravitational field is close to zero. Another way of looking at this increase of power is to remember that the distance fallen per second increases with speed. So potential energy lost per second increases with speed. So the rate at which gravity works increases with speed; which makes force × velocity more understandable.

Alternatively, a vessel filled with liquid and having a total thermal capacity of 5000 J K^{-1}, will be heated by a power supply of 800 W at a rate of (800 W)/(5000 J K^{-1}) or 0.16 K s^{-1}.

An electrical source transfers power to the external circuit at a rate of IV W where I is the current and V is the terminal potential difference.

$$\text{power} = \text{current} \times \text{voltage}$$
$$\text{watts} = \text{amps} \times \text{volts}$$

precision is a compound of the range of magnitudes that an instrument can measure and the smallest difference in magnitude that an instrument can resolve. The precision of a balance may be, say, ± 10 mg up to 200 g.

The problem with precision is to know when, and how far, it is significant. A micrometer screw gauge can be read to a precision of ± 0.01 mm up to about 3 cm without even trying to divide up the scale divisions by eye. With this kind of precision you can check the roundness of a thick copper wire. But you cannot measure the diameter of a piece of string since the jaws would squash the string.

Most instruments have a precision which goes beyond their reliability. A good voltmeter will be accurate to about ± 1% of full-scale deflection but the scale can be read more closely than this. A good magnifying glass will help read a temperature to ± 0.1 K from a mercury-in-glass thermometer which is reliable to no more than ± 1 K. In both cases the precision is swamped by systematic errors inherent to the instruments. (See *error, sensitivity, uncertainty*.)

prefixes used with SI units: Two sets of prefixes are used with SI units; a set of multiples and a set of sub-multiples. They are as follows:

multiples		sub-multiples	
deca–	× 10^1 da	deci–	× 10^{-1} d
hecto–	× 10^2 h	centi–	× 10^{-2} c
kilo–	× 10^3 k	milli–	× 10^{-3} m
mega–	× 10^6 M	micro–	× 10^{-6} μ
giga–	× 10^9 G	nano–	× 10^{-9} n
tera–	× 10^{12} T	pico–	× 10^{-12} p
peta–	× 10^{15} P	femto–	× 10^{-15} f
exa–	× 10^{18} E	atto–	× 10^{-18} a
zetta–	× 10^{21} Z	zepto–	× 10^{-21} z
yatta–	× 10^{24} Y	yocto–	× 10^{-24} y

pressure and its SI unit, the pascal (Pa), are defined by the word equations:

$$\text{pressure} = \frac{\text{thrust}}{\text{area}}$$

$$\text{pascal (Pa)} = \frac{\text{newton (N)}}{\text{metres squared (m}^2)}$$

Pressure is a non-directional *scalar quantity*. A solid can receive and transmit a *thrust* but it cannot transmit a pressure. Pressure is concerned with *fluids*, not *solids*. (These distinctions are never absolute. Some solids, especially polymers, behave like liquids when they are exposed to long-term forces.) The pressure in a fluid has a dynamic component related to its kinetic energy and a static component related to the weight of supported fluid and to any externally applied pressure; this is the basis of *Bernouilli's principle*. (See *liquid pressure (laws of), Archimedes' principle*.)

pressure law: this law states that the pressure of a fixed mass of gas at constant volume increases in proportion to the absolute temperature. It can be stated in algebraic form:

provided that m and V are constant, $p \propto T$

or $\dfrac{P_1}{T_1} = \dfrac{P_2}{T_2}$ – constant

Although the law is easily derived from the *ideal gas equation*, it is an experimental law and its validation by experiment was largely the work of the Frenchman Gay Lussac (1802). An account of a student experiment to verify the law should include a diagram, a list of the measurements and an explanation of how you would check the law. Notice that the apparatus used here is, in effect, a constant volume gas thermometer.

Experiment: to verify the pressure law

Apparatus

The apparatus is shown in the diagram.

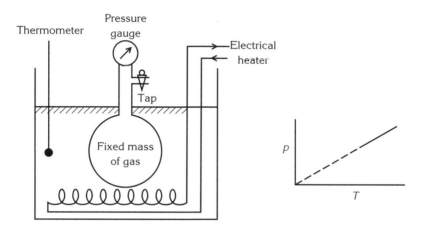

A fixed mass of dry air is trapped in a round flask which is immersed in a water bath. The pressure gauge has access to the air in the flask along a short narrow tube of negligible volume. An electric heater is used to control the temperature of the water. The mass of air and its volume are constant throughout the experiment.

Measurements

The atmospheric pressure is measured with a barometer. The temperature of the water and the pressure gauge reading are noted. The temperature of the water is increased by about 10 K, stirred and allowed to settle. The new temperature and the new pressure are noted once their values are constant. This procedure is repeated at about six temperature points up to about 80°C.

Conclusion

The temperatures are converted to the absolute scale T, by adding 273 K, and the gauge pressures are converted to gas pressures p by adding the atmospheric pressure. A graph is drawn of gas pressure p against absolute temperature T. If it is a straight line going through the origin, the pressure law is verified.

(See *ideal gas equation, Boyle's law, Charles' law, constant volume gas thermometer*.)

primary fuels are fuels such as coal, still in their raw state. (See *energy sources*.)

principle of moments states that, for a rigid body to be in equilibrium, the sum of the clockwise moments acting on the body must equal the sum of the anticlockwise moments acting on the body.

If this principle applies the body will not turn. But a second condition must be satisfied for the body to be in equilibrium: there must be no net force on the body in any one direction. (See *forces in equilibrium*, *moment of a force*.)

Worked example

The diagram shows a horizontal uniform steel bar AB of length 5 m and of mass 76 kg. It rests on two walls. Calculate (i) the smallest downwards force F_B at end B which is just strong enough to raise a bag of mass 24 kg hanging at the far end A and (ii) the force that the bar will then exert on the wall nearer to B. B is 1.20 m from the nearer wall.

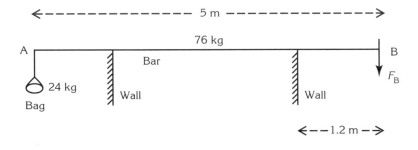

(i) Taking moments about the wall nearer to B:

Sum of the clockwise moments = sum of the anticlockwise moments

$$F_B \times (1.20 \text{ m}) = (76 \text{ kg})(9.8 \text{ m s}^{-2})(1.30 \text{ m}) + (24 \text{ kg})(9.8 \text{ m s}^{-2})(3.80 \text{ m})$$
$$= 968.24 + 893.76 = 1862 \text{ N m}$$
$$F_B = (1862 \text{ N m})/(1.20 \text{ m}) = 1.55 \text{ kN}$$

(ii) The downwards force on the wall is the sum of the three downwards forces the wall supports when the bar is about to tip, that is:

$$0.745 \text{ kN} + 0.235 \text{ kN} + 1.55 \text{ kN} = 2.53 \text{ kN}$$

principle of superposition states that the disturbance at a point where two wave motions meet is the vector sum of the disturbances from each of the two waves at that point. With two sound waves, the disturbances will be displacements in two different directions; the *parallelogram law* operates. With light, the right answers can generally be reached by adding the waves according to their amplitudes and their phase difference (or their path difference). This sounds all right but it leaves difficulties – how is light energy redirected from the dark parts of an interference pattern into the brighter parts? All too soon we are into quantum mechanics.

The principle of superposition is a statement of how two waves combine when they meet. Do not confuse it with the conditions for constructive and destructive *interference*.

progressive and stationary waves: see *stationary waves* for a contrast between these two phenomena.

progressive waves carry energy of oscillation through a medium if they are mechanical waves and through space if they are *electromagnetic waves*. (See *stationary waves, longitudinal waves, transverse waves, sound waves*.)

projectiles are objects, mostly solid, which are thrown upwards from the Earth's surface, mostly at an angle to the vertical, only to come down again. The situation is analysed in the usual way for *two-dimensional motion* and the *equations of motion*. The horizontal and vertical motions are tackled separately.

Worked example

A rubber ball is thrown to the ground at an angle and bounces 2.5 m high. It hits the ground again 4.0 m further on. If it loses 15% of the kinetic energy linked to its vertical motion at the second bounce, how far will it travel before the third bounce?

First, find the sideways velocity component from the time in the air for the first bounce.

Data: $s = 2.5$ m, $a = -9.8$ m s^{-2}, $v = 0$ m s^{-1} : Find t.

$$s = vt - \tfrac{1}{2}at^2$$
$$2.5 = 0 - \tfrac{1}{2}(-9.8 \text{ m s}^{-2})(t)^2$$
$$t = \sqrt{(5.0/9.8)} = 0.714 \text{ s}$$

The first bounce lasts for 2×0.714 seconds.

The sideways velocity component is displacement/time

$$= (4.0\text{m}) / (2 \times 0.714 \text{ s}) = 2.8 \text{ m s}^{-1} \quad \text{(to 2 s.f.)}$$

The kinetic energy per unit mass after a 2.5 m fall is gh, or 24.5 J kg^{-1}. 85% of this is 20.825 J kg^{-1}. This corresponds to $\tfrac{1}{2}v^2$, i.e., the vertical component of the velocity at the start of the second bounce is 6.45 m s^{-1}. Calculate the rise time for the second bounce:

Data: $u = 6.45$ m s^{-1}, $v = 0$, $a = -9.8$ m s^{-2}: find t.

$$v = u + at$$
$$t = \frac{v - u}{a} = \frac{-6.45}{-9.8}$$
$$= 0.66 \text{ s} \quad \text{(to 2 s.f.)}$$

The ball is in the air 1.32 seconds during the second bounce and has sideways velocity component 2.80 m s^{-1}. Calculate the length of the second bounce:

$$\text{displacement} = \text{sideways velocity component} \times \text{time}$$
$$= 2.8 \times 1.32 = 3.7 \text{ m} \quad \text{(to 2 s.f.)}$$

proton: an *elementary particle* with positive charge the same magnitude as the negative charge on the *electron* but whose mass is almost equal to the mass of a hydrogen atom. A hydrogen atom has just a proton for its nucleus.

proton decay and neutron decay are examples of the weak interaction at work. Both particles are baryons and their quark structures are (uud) for the proton and (udd) for the neutron. The up quark u has charge $+ \frac{2}{3}e$, the down quark d has charge $- \frac{1}{3}e$. The charge total for the proton is $+e$ while the charge total for the neutron is zero. When a neutron decays into a proton or vice versa, a beta particle is created to keep the charge in balance. But the beta particle is a lepton and an (anti)neutrino has to be created to conserve lepton number. It helps to tabulate the reactions and the quantum numbers that are conserved.

	neutron decay					proton decay				
	$n \rightarrow$	p^+	$+$ β^-	$+$	ν_e	$p \rightarrow$	n	$+$ β^+	$+$	ν_e
Baryon number B	$1 =$	1	$+$ 0	$+$	0	$1 =$	1	$+$ 0	$+$	0
Charge Q	$0 =$	$(+1)$	$+$ (-1)	$+$	0	$1 =$	0	$+$ 1	$+$	0
Lepton number L	$0 =$	0	$+$ $(+1)$	$+$	(-1)	$0 =$	0	$+$ (-1)	$+$	$(+1)$

These decay processes are illustrated in the entry *Feynman diagrams*. Those diagrams underline the fact that the beta particles and (anti)neutrinos are the decay products of the W exchange particles and that the exchange particles momentarily conserve the charge during the reactions.

proton number, Z, is another name for *atomic number*, the number of *protons* in the *nucleus*.

pseudo-vector quantities are quantities such as angular velocity which add together according to the parallelogram law but the directions of which are set by convention.

A body which is turned through a small angle experiences an *angular displacement*. If it is turned through another small angle about the same axis, then the total angular displacement is the sum of the two component displacements. If the two component angular displacements are about different axes, the addition is problematical. The answer is to represent the magnitude and direction of an angular displacement by a vector along the axis of rotation and satisfying the *right hand grip rule* (thumb of the right hand along the axis and fingers pointing in the direction of rotation). Because the direction of the angular displacement vector is wholly conventional, angular displacement is called a pseudo-vector. Angular velocity, angular momentum, angular acceleration and torque are pseudo-vectors in the same sense. None of this will be on your syllabus but you may come across the occasional reference to a pseudo-vector.

pulsar: see *neutron star*.

pulse code modulation (PCM) is the transmission of information in the form of a stream of equal-sized pulses or digital bits. An essential part of the process is *analogue-to-digital conversion* before transmission and *digital-to-analogue conversion* after transmission. Information which is transmitted in the form of digital bits can be cleaned of any noise during reception. This elimination of noise is a major advantage of PCM. Information is tied to the bit stream by a coding process. The bit stream might, for example, be broken up into a sequence of 12 bit groups, each consisting of two start bits, two stop bits and eight bits in the middle which can be coded in 256 different ways.

quantisation of charge means that electric charge is not a continuously variable quantity. The smallest possible increase or decrease in quantity of electric charge is the electronic charge $-e$ (charge on one electron). This was established by *Millikan's oil drop experiment*.

quantisation is an essential part of any *analogue-to-digital conversion* process. The voltage amplitude of the analogue signal at the input of a converter is a continuous variable. This infinite number of possible input values must now be shared among a fixed number of possible digital values at the output of the converter. It follows that each digital value corresponds to a range of analogue values. The small differences within this range can never be recovered. We say that the analogue values have been quantised. Quantisation always implies a small information loss.

quantity algebra is the practice of substituting into an equation numerical values with their units, all in SI. For example, to calculate the force F needed to induce an acceleration of 45 cm s^{-2} in a mass of 67 g, we would write:

$$F = ma = (0.067 \text{ kg})(0.45 \text{ m s}^{-2}) = 0.030 \text{ kg m s}^{-2} = 30 \text{ mN}$$

Quantity algebra is a discipline which minimises the chances of making simple errors in calculations, especially where multiples or sub-multiples of SI units are involved.

quantum mechanics is a term that covers those approaches to theories of fields and the structure of matter that hinge on the idea that the magnitude of a physical variable such as angular momentum must be one of a discrete set of values ($nh/2\pi$ for angular momentum). A consequence of this particular example is the quantisation of the orbital electron energies in an atom and a detailed explanation of atomic spectra.

quarks are the fundamental particles from which *hadrons* are constructed. Scattering experiments with 50 000 GeV electrons show them to be, like electrons, less than 10^{-18} m in diameter – that is, smaller than can be measured. Over 200 heavier particles discovered in cosmic rays or particle accelerators can be shown to be built up from six different kinds of quark and their antiquarks. These heavier particles are unstable and decay eventually into protons, neutrons, electrons and electron antineutrinos. Quarks are subject to all *four fundamental interactions*. The strong interaction holds quarks together in a *nucleon*, and nucleons together in a nucleus.

That quarks, unlike leptons, are subject to the strong interaction is attributed to a quality possessed by quarks known as colour. Colour is a bit like charge but it comes in three kinds, not two. The colours are called red, green and blue. The antiparticle quarks have anticolours. Quarks will form stable groupings only if the colours combine to form a neutral set. Either three quarks of three different colours (making white) form a baryon or three antiquarks with three different anticolours form an antibaryon or two opposite colours (red and anti-red to make colouress) form a meson. The main properties of the six known kinds of quark are listed in the following table:

Flavour	Generation	Mass/MeV	Charge/e	Baryon no.	S	C	B	T
d down	1	5–15	−1/3	1/3				
u up	1	2–8	+2/3	1/3				
s strange	2	100–300	−1/3	1/3	−1			
c charm	2	1000–1600	+2/3	1/3		1		
b bottom	3	4100–4500	−1/3	1/3			−1	
t top	3	180 000	+2/3	1/3				1

Flavour is a little more than six labels. Flavour is conserved in strong interactions. The four quantum numbers S, C, B, T take opposite values in the antiparticles and are involved in the conservation of the individual flavours. Flavour is not conserved in processes controlled by the weak interaction (see *proton decay and neutron decay*). Antiquarks have opposite values of charge and baryon number to those listed in the table. All quarks have spin $\pm\frac{1}{2}$ and these spins combine to give $\pm\frac{3}{2}$ or $\pm\frac{1}{2}$ for baryons and ±1 or 0 for mesons.

The strong interaction responsible for bonding quarks together in hadrons is associated with eight different *exchange particles* or gauge bosons called gluons. A gluon carries both colour and anticolour; it can bind to another gluon or bind two quarks together. The bonding is so strong that the energy needed to pull a quark out of a nucleon is always sufficient to create a second meson or boson. Quarks cannot be isolated.

A proton has two up quarks and one down quark (uud), giving a charge $+e$ and baryon number 1. The antiproton has charge $−e$ and baryon number −1. The neutron (udd) has charge zero and baryon number 1. *Mesons* consist of a quark and an antiquark, so baryon number is always 0, but because the flavours are different the charge may be $+e$, $−e$ or 0.

quasars (quasi-stellar radio sources) are stellar objects which share four characteristics: a large red shift which implies that their distances are greater than those of remote galaxies, luminosities (as calculated using the inverse square law) in excess of many galaxies, a small size comparable with the solar system; a varied radiation output which includes visible light, radio waves and X-ray radiation all varying in intensity over a time scale of about a year. They remain a puzzle.

quenching is the immersion of a hot metal in cold water. The rapidity of the cooling process limits or prevents the *annealing* process. It makes the metal hard and brittle. (See *annealing, tempering.*)

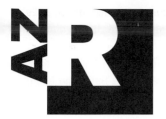

radial electric field: the electric field near an isolated point electric charge. The diagram below shows the *field lines* and the equipotential lines (broken lines) near an isolated positive charge. Notice how the field lines get further apart as the field weakens. Notice, also, how the spacing of the equipotentials increases if the change in potential from line to line is always the same.

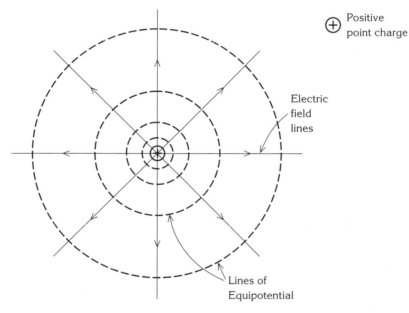

radial magnetic field: the magnetic field in the space between the poles of a correctly-shaped horseshoe magnet and a soft iron cylinder placed symmetrically between the poles. The magnetic field within the space is everywhere aligned with the direction from the centre of the cylinder.

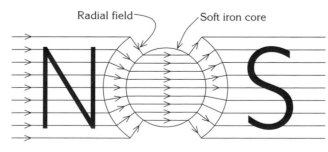

A coil suspended within the radial field is effectively immersed in a magnetic field parallel to the plane of the coil. In these circumstances, if the coil carries current, the *torque* on the coil has a maximum value for that current. (See *electric motor*.)

radian is a unit of angle. One radian is the angle subtended at the centre of a circle by an arc of length one radius. The symbol for radian is always written 'rad'. Since the length of the circumference of a circle is $2\pi \times$ radius, 360° is 2π radians.

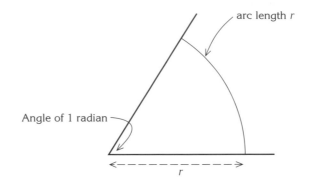

arc length r

Angle of 1 radian

r

radiation is energy travelling through space in the form of *photons*. (See *electromagnetic spectrum*.)

radiation detectors are broadly of two kinds; those which are used for ionising radiations (mostly alpha, beta and gamma radiation, X-rays and sometimes UV) and those which are used to observe electromagnetic radiation only, from UV down to radio waves.

Ionising radiations are detected with photography, GM tubes, cloud chambers, bubble chambers, ionisation chambers, scintillation counters, gamma cameras and solid state detectors. Longer wave electromagnetic radiation is detected, as well as by photography, by bolometers, thermopiles, *light-dependent resistors* and, for longer wavelengths, aerials.

GM tubes consist of a conducting cylinder insulated from a conducting needle which runs down its axis. One end is closed by a thin mica window and the other end by the insulating material. The cylinder is filled with argon and a trace of chlorine. A potential difference of a few hundred volts is applied across the cylinder and needle arrangement and some circuitry. Any ionising particle which enters the tube ionises the gas. The sudden flow of charge operates a counter and switches off the power to the tube for a time interval long enough for the tube to recover (a few microseconds). The tube can be used to detect alpha and beta particles individually. The count rates for gamma radiation and X-rays are in proportion to the radiation flux. With high count rates, a correction has to be made for the length of time the tube is out of action (i.e. the dead time).

Cloud chambers are vessels filled with air saturated with water vapour. Sudden expansion causes cooling of the air and its supersaturation. Any ionising particle which passes through the air at this instant will trigger condensation along its path, and this condensation track can be photographed. The great advantages are that the track density identifies the particle, track curvature identifies the charge and track length identifies the energy. Bubble chambers are filled with liquid hydrogen instead of air and the bubbles that arise result by.boiling, not from a condensation process. The advantages this time are the ability to track much more energetic particles and to investigate particle–proton interactions.

Ionisation chambers work in a similar way to GM tubes but they don't switch off at each detection. They use higher voltages and deliver a current which is a measure of ionisation rate. The ionisation rate, in its turn, is an indicator of ionising particle flux.

A *scintillation counter* consists of a sodium iodide scintillator crystal, laced with thallium, NaI(Tl), in optical contact with a photomultiplier. Any ionising particle which enters the crystal is absorbed and its energy mostly re-appears as a burst of visible photons. Many of these photons reach the photocathode of the photomultiplier and release a number of photoelectrons. These photoelectrons make their way up a ladder of about a dozen dynodes in the photomultiplier powered by a few hundred volts. Each electron releases several other electrons at each dynode reflection. The outcome is a burst of current at the anode of photomultiplier. The height and width of this pulse measures the energy released by the ionising particle. The scintillator counter assembly is calibrated against individual particles of known energy. The whole arrangement is enclosed in an opaque case because photomultiplier tubes are very sensitive to light.

Solid state detectors are mostly p–n junctions under reverse bias. A particle absorbed at the junction will release a burst of electron–hole pairs. The height and width of the current pulse are indicators of incident particle energy. Calibration is necessary.

Bolometers are thin strips of blackened platinum. Placed across a low intensity beam of radiation, the bolometer strip warms up to an equilibrium temperature and its electrical resistance increases. The higher electrical resistance is measured. Next, away from the radiation, a small current is passed through the bolometer. This current is increased until the bolometer resistance is the same as when it was irradiated. The electrical power dissipated in the bolometer, I^2R, equals the radiation intensity which warmed up the bolometer strip.

Thermopiles are constructed from a large number of *thermocouples* in series. The incident radiation warms up the hot junctions and incident radiation intensity is indicated by the c.m.f. generated. Light-dependent resistors are semiconducting resistors in which there is an increased density of current carriers when the resistor is illuminated. Longer wavelengths, from microwaves to radio waves, are detected with aerials.

radiation dosage involves four related ideas: absorbed dose, quality factor, the dose equivalent and exposure. Absorbed dose is the beam energy absorbed per kilogram of body tissue. The unit, the gray (Gy), is an absorbed dose of 1 J kg^{-1}.

Quality factor is a number with no unit. It is defined as the ratio:

$$\text{quality factor} = \frac{\text{absorbed dose using 250 kV X-radiation}}{\text{absorbed dose of radiation having the same effect}}$$

Dose equivalent in sieverts (Sv) equals absorbed dose × quality factor. The total dose equivalent for a patient subjected to different types of radiation is the sum of the individual dose equivalents.

The fourth idea, 'exposure', is an amount of X-radiation or gamma radiation measured by the total negative charge produced by ionisation in a kilogram of air. The unit is C kg^{-1}. It can be measured without danger to the patient. The product of exposure in C kg^{-1} and a conversion factor *f* gives the absorbed dose in Gy. The conversion factor *f* is a known number, differing for each tissue type.

Radiation damage that is certain to happen, such as tissue destruction, is non-stochastic. Damage such as mutation, which may or may not happen, is stochastic. (Stochastic is another word for random.)

radiation loss is the energy radiated into the surrounding space by a hot object. (See *Stefan's law*.)

radio waves are *electromagnetic waves* with wavelengths longer than about ten centimetres. The modest frequencies (MHz or lower) have limited information-carrying capacity and this has led to the 'crowding of the air waves'.

radioactive decay is the spontaneous but delayed emission of an alpha or beta particle (positive or negative) from a *nucleus* . The atomic number (proton number) of the nucleus must change. *Gamma emission* is a way of ridding the nucleus of excess energy when it is in an excited state following a radioactive decay event. Gamma emission is not radioactive decay. (See *alpha decay*, *beta decay*.)

radioisotopes are found in nature only if they have long half-lives. Short-lived radioisotopes are mostly prepared by irradiation techniques. Examples include alpha particle bombardment [e.g. $^{121}Sb(\alpha,2n)^{123}I$] and neutron bombardment [e.g. $^{130}Te(n,\gamma)^{131}Te(\beta^-)^{131}I$]. Where the half-life is shorter than a day or so, radioisotopes are prepared close to the point of use. An example is the preparation of the metastable isotope ^{99m}Tc (half-life six hours) which is a decay product of ^{99}Mo (half- life almost three days) and is separated from the molybdenum by washing the molybdenum with saline solution. The saline dissolves the pertechnetate ions but not the molbdenate ions.

Radioisotopes may be used for diagnostic purposes (usually an imaging process) or for therapy (usually a destructive process). Imaging processes employ gamma emitters; the gamma radiation is detected outside the body with a *gamma camera* or photographically. The energy of the individual gamma photon is generally not far from 150–250 keV.

Therapy may involve introducing a beta emitter which irradiates a small volume of tissue centred on the source. Beta particle energy is often ten times as much as gamma photon energy, around one MeV. This, together with the much higher attenuation coefficient, accounts for the damage sustained by nearby tissues. An alternative is to use a cobalt-60 source of high energy gamma photons (1.17 MeV and 1.33 MeV) in which case the rest of the body is protected by screening.

Radioisotopes absorbed into the body can be removed either by the natural decay process (half-life t_r) or by a biological rejection process (half life t_b). These two processes combine to give an effective half life t_e given by

$$\frac{1}{t_e} = \frac{1}{t_r} + \frac{1}{t_b}$$

Radioisotopes used in medicine must be pharmaceutically pure and collect in the organ or system they will be used to treat. This may be achieved by using the blood system or by direct injection.

random error is a discrepancy, which comes about by accident, between the true value of a physical quantity and the actual value of the measurement of the physical quantity. A digital clock never stops or starts at exactly the right time, a scale might be read slightly on the low side or slightly on the high side. *Parallax error* is a common example of a random error. There is no way of knowing the size of the random error present in a single measurement since the true value will not be known. But random errors can be reduced by taking the average of several independent measurements. This is possible because a random error is as likely to be positive as negative, to be larger or smaller. (If you study statistics, then you will notice that the different measurements must be normally distributed about the true value unless there is a systematic error.) Random errors, like *systematic errors*, contribute to the *uncertainty* of a measurement.

random motion is the motion of an atom in a fluid whose velocity changes randomly in both magnitude and direction as a consequence of repeated collisions. The distance travelled between collisions is also random. A change is random if the new value cannot be determined from previous values. (Compare with *Brownian motion*.)

randomness and radioactive decay are two concepts which are linked by the laws of probability. An event is random if the probability of its happening in a given time interval depends on the length of the time interval but is independent of when the time interval begins (i.e. independent of age). This idea can be stated mathematically and applied to radioactive decay.

Let λ be the probability per second that a nucleus will decay. Then the probability that a nucleus will decay during a time interval of length Δt seconds is $\lambda.\Delta t$.

If there are N nuclei waiting to decay at some time t, then the expected number of nuclei, ΔN, which will decay between times t and $t + \Delta t$ is $(N \times \lambda.\Delta t)$. That is:

$$\Delta N = -N.\lambda.\Delta t = -\lambda N.\Delta t$$

This equation is called the *fundamental law of radioactive decay*. The negative sign tells us that ΔN is a reduction in the value of N. The probability λ is defined by the above equation. It is the probability per second that a nucleus will decay. It is called the decay constant; its unit is second^{-1}. See *decay constant*, *exponential law of radioactive decay*.

rate of change of a physical quantity is the increase per unit time in the magnitude of the physical quantity. This increase per unit time of the temperature T, shown in the graph can be written in either of two ways: $\Delta I/\Delta t$ or dI/dt. The difference between these two ways is of some importance and can be understood from the diagram. The diagram is a graph of the temperature T of a saucepan against time t. The increment in time, $(t_B - t_A)$, is written Δt. The corresponding rise in temperature, $(T_B - T_A)$ is written ΔT.

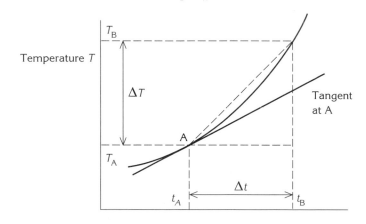

Time t

The average rate of change of temperature during the time interval t_A to t_B is written $\Delta T/\Delta t$. The *limiting value* of this average rate of change as the time interval Δt is made progressively smaller until it approaches zero is written dT/dt. This is the instantaneous value of the rate of change.

Δt implies a small but finite time increment, dt represents an infinitesimal time increment. Rate of change is positive for a growth and negative for a decay. dT/dt at time t_A is the gradient of the tangent to the curve at the point A.

The x-axis does not have to be a time scale. Using more formal language, the rate of change of a dependent physical variable y with respect to an independent physical variable x is the gradient of the tangent to the graph of y against x for the relevant value of x. See *limiting value*.

rate of cooling means rate of fall of temperature. It does not mean rate of loss of heat. It equals the gradient of a temperature against time graph at the relevant instant. (See *Newton's law of cooling*.)

ray diagram: a diagram of an optical lens or mirror system which accounts for the formation of the image. A ray diagram accounts for the formation of a real image by showing the rays of light diverging from the object position, meeting again at the image position and thereafter diverging away from the image position. For a virtual image, the rays diverge from a point behind the lens or mirror. (See *lens equation*, *plane mirror*, *ray optics*.)

ray optics is an analysis of mirrors and lenses based on the idea that light travels in straight lines. It is successful except where the outcome of an optical arrangement hinges on the wave nature of light.

reactance is the ratio of peak voltage to peak current for a capacitor or for the inductive component of an inductor (how the inductor would behave if it had zero ohmic resistance). Reactance is $1/\omega C$ for a capacitor and ωL for a pure inductor. The SI unit for reacitance is the ohm (Ω). Reactance differs from resistance in that the voltage and current for a capacitor or for a pure inductor are not in phase, that is, they peak at different times. (See *capacitive reactance*, *series resistance and capacitance*, *inductive reactance*, *series resistance and inductance*.)

rectification, half-wave and full-wave, are methods for changing an alternating current into a direct current. Remember that 'alternating' means 'changing direction' and not 'changing magnitude'.

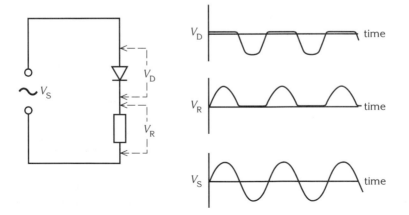

The half-wave rectifier in the first circuit, consists of an alternating power supply in series with a diode and resistor. Assuming that the diode has infinite resistance for current in one direction and zero resistance to current in the other direction, the circuit will conduct in alternate half-cycles. The input voltage falls across the diode in the non-conducting half-cycles and across the resistor during the conducting half- cycles. The three graphs in the first diagram show how these two voltages add up to the supply voltage.

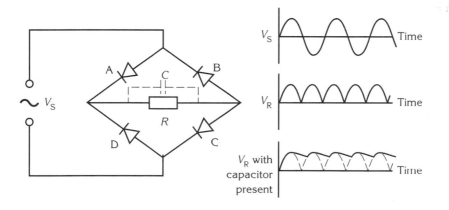

The full-wave rectifying circuit shown in the second diagram employs four diodes in a bridge circuit. Whatever the direction of the voltage from the AC power supply, current flow through the resistor at the centre of the bridge from left to right. It helps to remember that the four diodes should all point towards the left (or all point towards the right, in which case the current in the resistor flows the other way). During positive half-cycles, A and C conduct. During negative half-cycles, D and B conduct. The second graph shows the pulsating but direct voltage across the resistor. This is the output voltage and the resistor represents the load. A large capacitor connected in parallel with the load smoothes the output voltage.

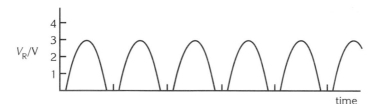

The third graph shows the effect on the output voltage of connecting a capacitor across the load. The capacitor charges up when the input voltage is growing and discharges slowly when the input voltage falls away. After a few cycles the circuit reaches the situation indicated in the graph with a small ripple added to a steady DC voltage. The ripple amplitude increases if more current is drawn for the load resistor.

The two circuit diagrams are drawn on the assumption that the diode is perfect. In practice diodes need a voltage drop of half a volt or so in the forward direction. If, therefore, four diodes are used to fully rectify a low voltage supply (say 4 V at the peak) the output voltage will include a gap between the pulses.

red giant: the first stage in the decay of a main sequence *star* whose mass is less than about eight solar masses. Although light output during this stage is much higher than during the main sequence stage, the proportionate increase in surface area is even greater. Less radiation is emitted per unit surface area, surface temperature falls and the peak wavelength shifts towards the red. See *end states of stars*, *Wien's law*.

reflection of light: the change of direction of a light ray at the surface of a medium which ensures that the light ray remains in the one medium. The change of direction obeys two laws known as the laws of reflection.

1st law   The incident ray, the reflected ray and the normal to the reflecting surface at the point of incidence, all lie in the same plane.

2nd law The angle of incidence equals the angle of reflection.

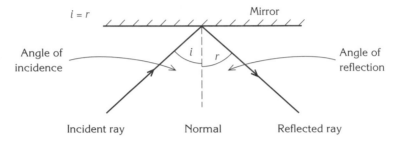

The first law states that reflection is a two-dimensional phenomenon which can be represented by a two-dimensional diagram on a flat piece of paper. The second states the principle to be followed when drawing these diagrams. (See *plane mirrors*.)

refraction of light is the bending of light when it passes from one transparent medium into another. The change of direction obeys two laws known as the laws of refraction. These are:

1   The incident ray, the refracted ray and the normal to the refracting surface at the point of incidence all lie in the same plane.

2   The sine of the angle in incidence divided by the sine of the angle of refraction is a constant for two given media.

The diagram illustrates a typical situation.

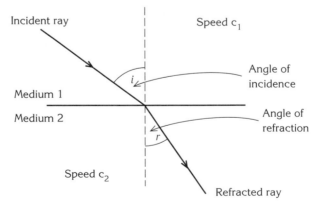

The first law states that refraction is a two-dimensional happening which can be represented by a two-dimensional diagram on a flat piece of paper. The second states the rule which enables us to trace the path of a ray of light from one medium into another. The constant which is defined by the second law of refraction is called refractive index. With this in mind, the second law can be re-written in the form:

$$\frac{\text{sine of the angle of incidence}}{\text{sine of the angle of refraction}} = \text{refractive index}$$

$$\frac{\sin i}{\sin r} = n$$

This last equation is known as Snell's law; it defines refractive index. But electromagnetic theory gives us another relationship:

$$\text{refractive index} = \frac{\text{speed of light in the incident medium}}{\text{speed of light in the refracting medium}} = \frac{c_1}{c_2}$$

It is usual practice to identify the materials and the direction with a subscripted prefix for the material on the incident side and a subscripted suffix for the material on the far side. For light travelling from air to glass, for example:

$$_{air}n_{glass} = \frac{\sin i}{\sin r} = 1.52 = \frac{\text{speed of light in air}}{\text{speed of light in glass}} = \frac{c_{air}}{c_{glass}}$$

1.52 is the refractive index of ordinary glass (called crown glass). Heavier flint glasses have higher refractive indices. For light travelling from glass into air:

$$_{glass}n_{air} = \frac{\sin r}{\sin i} = \frac{1}{1.52} = \frac{\text{speed of light in glass}}{\text{speed of light in air}} = \frac{c_{glass}}{c_{air}}$$

These statements can be generalised for any two transparent materials:

$$_1n_2 = \frac{\sin \theta_1}{\sin \theta_2} = \frac{c_1}{c_2} = \frac{1}{_2n_1}$$

See *refraction of light, total internal reflection, critical angle.*

refrigerator: a device which removes internal energy from a low temperature container and discards it to the surroundings at a higher temperature. It is a *heat engine* worked backwards; work W is done on the refrigerator, internal energy Q_2 is removed from the container and energy $(W + Q_2) = Q_1$ is transferred to the surroundings.

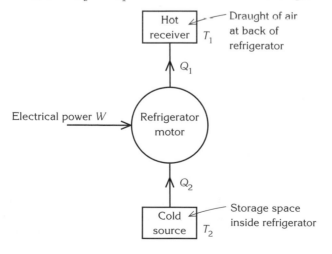

It follows from the theory of the heat engine that:

$$\frac{W}{Q_1} = \frac{Q_1 - Q_2}{Q_1} \left\{ = \frac{T_1 - T_2}{T_1} \right\}$$

If the heat engine being worked backwards is not a perfect engine, the part of the expression involving absolute temperatures (in large brackets) fails.

The purpose of a refrigerator is to reduce the internal energy of the container. The coefficient of performance for a refrigerator is defined with this purpose in view:

$$\frac{Q_2}{W} = \frac{Q_2}{Q_1 - Q_2} \left\{ = \frac{T_2}{T_1 - T_2} \right\}$$

As before, if the engine of the refrigerator is less than perfect, the part of the expression in large brackets fails. (See *heat engine concept*, *heat pump*.)

reinforced concrete is concrete which is cast and set over a number of steel rods to give a beam which will withstand greater bending forces than concrete would on its own. It should not be confused with pre-stressed concrete. (See *composite materials*.)

relative atomic mass is the number of times the atom of a nuclide is heavier than one-twelfth of the mass of a carbon-12 atom. (A nuclide is a substance whose atomic and mass numbers are specified.) The word equation definition is:

$$\text{relative atomic mass of a nuclide} = \frac{\text{mass of an atom of the nuclide}}{\text{mass of an atom of } ^{12}_{6}C} \times 12$$

Alternatively, the relative atomic mass of a nuclide is numerically equal to how many times the mass of an atom of the nuclide is greater than the *unified atomic mass constant*.

relative molar mass of a substance is defined by the word equation:

$$\text{relative molar mass of a substance} = \frac{\text{mass of one mole of the substance}}{\text{mass of one mole of } ^{12}_{6}C} \times 12$$

(See *mole*.)

relative permeability μ_r is the ratio of the *magnetic field strength B* within a material to the magnetic field strength B_0 of the magnetising field.

$$\mu_r = \frac{B}{B_0}$$

It is slightly less than unity for *diamagnetic* substances, slightly greater than unity for *paramagnetic* substances and large and positive for *ferromagnetic* substances. (See *Curie temperature*, *solenoid*.)

relative permittivity of an insulating substance is the factor by which the *capacitance* of a vacuum-filled capacitor would increase if the space between the plates were to be filled by the insulating substance. It is defined for a particular material by the following word equation:

$$\text{relative permittivity } \varepsilon_r = \frac{\text{capacitance of a capacitor filled with the material}}{\text{capacitance of the same capacitor filled with vacuum}}$$

ε_r is defined as a ratio and, as a consequence, has no unit.

Most dielectric materials have relative permittivities less than 10. The success of a dielectric material in increasing the range of capacitor values while keeping the overall dimensions of the capacitor small is mainly because the dielectric material can be manufactured in the form of a very thin sheet. The conducting plates can then be held a very small distance apart without touching.

relativity: see *special relativity*, *general relativity*.

renewable energy sources are *energy sources* such as solar energy, *biofuels*, wind power, tidal power or hydroelectric power, which draw their power from the recent activity of the sun or geothermal power, which are not finite resources in the same sense as something like coal. (See *primary fuels*, *finite energy sources*.)

resilience is the capacity of an elastic material to withstand repeated stress and relaxation without deterioration despite appreciable elastic *hysteresis* effects. Car tyres are resilient. (See *rubber*, *elasticity*.)

resistance is the ratio of the *potential difference* (voltage) across a conductor to the *current* it carries. It is defined by the word equation:

$$\text{resistance} = \frac{\text{potential difference}}{\text{current}}$$

$$\text{ohm} = \frac{\text{volt}}{\text{amp}}$$

The symbol used is R and the SI unit is the ohm (Ω). Any circuit component or circuit, however complex, provided that it carries a stable current for a particular voltage across it, has resistance calculated from the definition in the word equation above. *Ohm's law* only applies to some conductors, those which obey it. (See *resistivity*, *temperature coefficient of resistance*, *internal resistance*.)

resistance thermometer: a thermometer which relies on the variation of electrical resistance with temperature. Platinum resistance thermometers are extremely accurate but expensive, bulky and need a lot of care. They are generally used as so-called secondary standards (or, even worse, as sub-standards) for calibrating other thermometers.

resistivity, for a material which is available in the form of a wire, is defined by the word equation:

$$\text{resistivity } (\rho) = \text{resistance } (R) \times \frac{\text{area of cross section } (A)}{\text{length } (l)}$$

$$\rho = R \times \frac{A}{l}$$

$$\Omega\,\text{m} = \Omega \times \frac{\text{m}^2}{\text{m}}$$

Resistance is a property of a circuit component, typically a wire. Resistivity is a property of the material the component is made from, typically a metal. The graph on page 206 show the resistivity of tungsten wire at different *absolute temperatures*.

(Note that you would lose all the marks if, in a sentence definition of resistivity, you wrote '... per unit area of cross-section per unit length ...'. as area is above the line and is not a 'per'. The word equation is unambiguous, easier and quicker.)

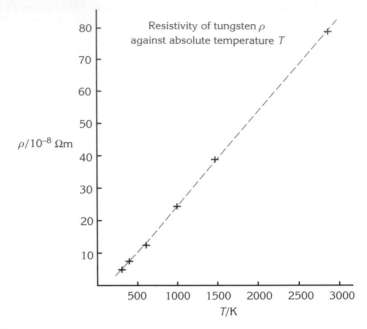

Resistivity of tungsten ρ
against absolute temperature T

$\rho/10^{-8}\ \Omega m$

T/K

Resistivity

resistors in parallel are resistors which are all connected across the same two points as shown in the diagram below. We need to know how to replace these three resistors with a single resistor R without changing either the current total I drawn from the rest of the circuit or the potential difference V across the arrangement.

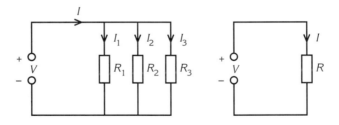

The current I in R must equal the sum of the other three currents in the three resistors:

$$I = I_1 + I_2 + I_3$$

Applying Ohm's law to this condition:

$$\frac{V}{R} = \frac{V}{R_1} + \frac{V}{R_2} + \frac{V}{R_3}$$

The voltage V cancels through and we are left with:

$$\frac{1}{R} = \frac{1}{R_1} + \frac{1}{R_2} + \frac{1}{R_3}$$

Worked example: Find the value of the unknown resistance *R* in the circuit shown.

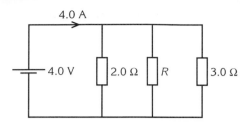

The equivalent resistance of all three resistors equals (4 V) / (4.0 A) = 1 Ω, i.e.,

$$\frac{1}{1} = \frac{1}{2} + \frac{1}{R} + \frac{1}{3} = \frac{5}{6} + \frac{1}{R}$$

$$\frac{1}{6} = \frac{1}{R} \quad \text{and} \quad R = 6\,\Omega$$

resistors in series are resistors which are connected end to end as shown in the diagram below. We need to know how to replace the three resistors with a single resistor *R* without changing either the current *I* drawn from the rest of the circuit or the potential difference *V* across the arrangement.

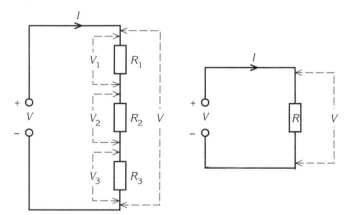

The voltage *V* across *R* must equal the sum of the other three voltages:

$$V = V_1 + V_2 + V_3$$

Apply Ohm's law to this condition:

$$IR = I_1 R_1 + I_2 R_2 + I_3 R_3$$

Kirchhoff's first law states that:

$$I_1 = I_2 = I_3$$

and all three currents equal *I* if resistance *R* is to be equivalent to the other three. The currents cancel through and we are left with:

$$R = R_1 + R_2 + R_3$$

Worked example: Find the value of the unknown resistance R in the circuit shown below.

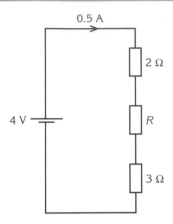

The equivalent resistance of all three resistors equals (4 V) / (0.5 A) = 8 Ω, i.e., 2 Ω + R + 3 Ω = 8 Ω, and R = 3 Ω.

resolution of vectors usually means finding the horizontal and vertical components of a single oblique vector. The diagram below shows a velocity vector v inclined at 55° to the ground. The vertical and horizontal directions are given. Dropping two perpendiculars from the far end of the vector constructs a rectangular 'parallelogram'. The horizontal component, v cos 55°, and the vertical component, v sin 55°, can both be read from the diagram.

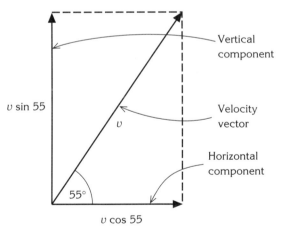

Resolving displacement and velocity vectors in this way simplifies a problem where the vertical motion is subject to gravitational force and acceleration while the horizontal motion continues at constant velocity.

A set of coplanar forces will be in equilibrium provided that the components of all the forces in each of two directions at right angles add up to zero and provided that the sum of the moments of the forces about any point in the plane is zero. (See *forces in equilibrium*.)

resolving power is a measure of the fineness of detail distinguishable in the focused image formed by an optical instrument or by something similar, such as a radio telescope.

The definition has a simple mathematical form illustrated in the diagram. The resolving power of an image-forming instrument is the minimum angular distance in radians, measured at the instrument, of two small objects in the field of view which can just be distinguished as two separate objects in the image.

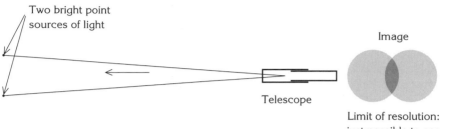

Two bright point sources of light

Telescope

Image

Limit of resolution: just possible to see that there are two point sources

Resolving power is limited by diffraction effects in the instrument. For an optical or radio telescope,

$$\text{resolving power } \theta \text{ in radians } = \frac{1.22\lambda}{d}$$

where λ is the wavelength of the light or the radio waves focused by the instrument and d is the diameter of the circular hole which limits the width of the beam of radiation entering the instrument. This diameter d is called the aperture. The equation for resolving power explains why the diameters of the dishes of radio telescopes have to be so large, that is, why they need so large an aperture. Radio waves have long wavelengths and θ needs to be small.

resonance is the setting up of large amplitude oscillations in a damped oscillator when the driver frequency matches the natural frequency of the oscillator. This frequency is then called the resonant frequency. The diagram below shows the change of equilibrium amplitude with driver frequency for the case when driver frequency changes, but driver amplitude is constant. Also shown is the effect of heavy damping contrasted with light damping. The finite amplitude at zero frequency corresponds to the displacement of the system under the action of a constant displacing force.

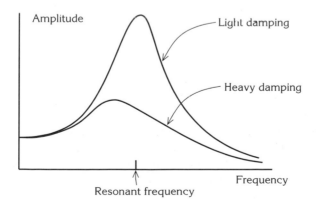

Amplitude

Light damping

Heavy damping

Frequency

Resonant frequency

To understand the phenomenon, imagine you are pushing a child on a swing and you want to build up the amplitude with minimum effort. You will find yourself applying a force in step with the displacement and in the direction of motion. The effect of this is to ensure that your driving force matches the period of the swing and the periodic damping forces. Also, by keeping in step with the direction of motion, you ensure that the applied force is always enhancing the motion. Change slightly the frequency of the pushes you apply to the swing and you will find that, after a little while, you are so out of step that your pushes are opposing the motion of the swing and slowing it down. The more you change the frequency the worse this situation gets. The only way to succeed is to match the push frequency to the system it is pushing, and this is resonance. (See *forced oscillations*.)

resonance in AC circuits is of two kinds. The impedance of a series circuit has a minimum value at the resonant frequency, the impedance of a parallel circuit has a maximum value at the resonant frequency.

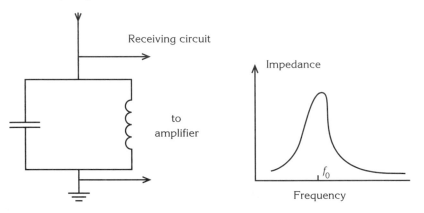

The parallel circuit shown in the diagram above is set up as a tuning circuit. The aerial picks up radio signals of many frequencies. The inductor has impedance ωL; it offers a low impedance pathway to earth to the lower frequency signals. The capacitor has impedance $1/\omega C$; it offers a low impedance pathway to earth to the higher frequency signals. The impedance of the parallel pair is a maximum at the angular frequency for which the two impedances are equal:

$$\omega L = \frac{1}{\omega C} \quad \text{and} \quad \omega = \frac{1}{\sqrt{(LC)}}$$

This will be the carrier frequency for the radio signal which sets up the maximum amplitude voltage oscillation across the parallel pair and which is passed to the next stage for amplification.

The series resonant circuit is shown on page 211, along with the phasor diagram.

Should the value of the angular frequency be such that the voltage across the inductor is equal to, and in the opposite direction to, the voltage across the capacitor, then the supply voltage will be wholly taken up with driving the current in the resistor and the current amplitude will have its maximum value. Once again, this happens at the frequency for which:

$$\omega L = \frac{1}{\omega C} \quad \text{and} \quad \omega = \frac{1}{\sqrt{(LC)}}$$

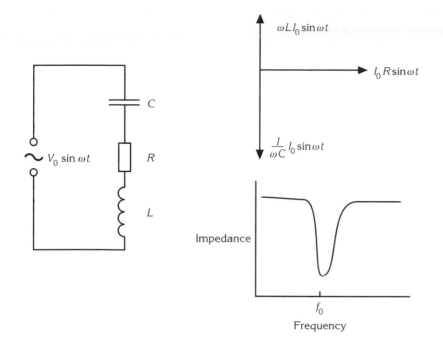

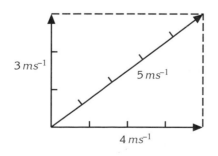

The condition of resonance needs many complete cycles to be properly set up, but since the frequencies are usually measured in kHz, this takes only a short time. Once the resonant condition is achieved, energy oscillates between the electric field in the capacitor and the magnetic field in the inductor, carried by the current. The power supply provides the electrical energy dissipation in the resistor. (See *reactance, capacitive reactance, series resistance and capacitance, inductive reactance, series resistance and inductance.*)

resonance with sound waves: See *waves on strings, waves in pipes.*

resultant is the name given to the single vector which equals the sum of two or more other vectors. The diagram below shows the resultant velocity (5 m s^{-1}) of two velocities at right angles, 3 m s^{-1} and 4 m s^{-1}. (See *addition of vectors, parallelogram law.*)

Reynold's number, R, equals $\rho vD/\eta$ where ρ is the density of a fluid, v is the speed of the fluid through a passageway or past an obstacle, D is a typical dimension (the width) of the passageway or the obstacle and η is the coefficient of viscosity of the fluid. If R exceeds about 2500 (in SI units), then *streamline flow* is impossible and *turbulent flow* ensues.

right hand grip rule: a rule for finding the direction of the magnetic field near a long straight wire or within a *solenoid*. For the straight wire, grasp the wire with the right hand, with the thumb pointing in the direction of the current. The fingers now point in the direction of the magnetic field. For the solenoid, grasp the outside of the solenoid with the right hand and with the fingers following the current direction in the wire; the thumb will be pointing in the direction of the magnetic field *within* the solenoid. Either of these methods can be used for the current ring. (See *ring, solenoid, straight wire, Maxwell's corkscrew rule.*)

rigid bodies are imaginary bodies which do not change shape or distort in any way when acted upon by a number of forces. The idea allows the mathematician or physicist to concentrate on the external effects of applying a set of forces to a body. (See *rigidity.*)

rigidity is the ability of a solid body to retain its shape when placed under *stress.*

ring: one or several circular turns of a single wire bound together in a single circular loop of radius r. If there are N turns and if current I flows in the wire then the *magnetic field strength* B at the centre of the ring is given by

$$B = \frac{\mu_0 NI}{2r}$$

The shape of the magnetic field is shown in the diagram. The direction of the field can be found from the right hand grip rule (imagine that the ring is a short solenoid).

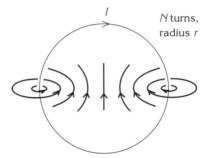

ripple tank: a tray, about 50 cm across, with a transparent base and capable of holding water to a depth of about one centimetre. Ripples are set up on the surface of the water with a vibrator. The vibrator may be a thin vertical rod just touching the surface of the water or it may be long and straight. The thin vertical rod sends out circular ripples, the straight rod sends a stream of straight ripples across the tank. For depths of less than a centimetre and for frequencies less than about 20 Hz, the speed of the ripples increases from about 0.15 m s^{-1} to about 0.25 m s^{-1} with increasing water depth and vibrator frequency.

The wavelength of the ripples is varied by changing the frequency of the vibrator. These ripples are used to demonstrate the meaning of words such as *wavelength, diffraction* and *interference,* used when discussing the properties of waves generally.

Two important demonstrations are illustrated in the diagram above. Straight waves arriving at a barrier with a small gap will diverge outwards after penetrating the gap. This is called diffraction. The smaller the gap, the stronger the diffraction. Where two sets of circular ripples overlap, a new and more complex pattern (interference) is set up; the rule which controls the way the two wave patterns combine is called the *principle of superposition.*

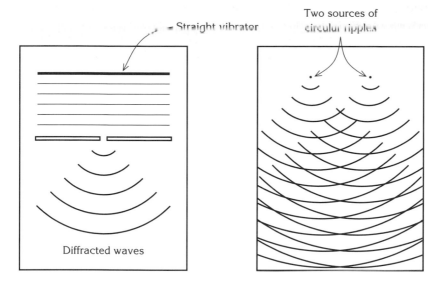

Two sources of
circular ripples

Straight vibrator

Diffracted waves

Ripple tank

r.m.s. (root mean square) values of alternating currents and voltages are the direct currents and voltages which would have the same power dissipation in a resistive load. If I_0 and V_0 represent the peak values of the alternating current and voltage, and if I_{rms} and V_{rms} represent the root mean square values:

$$I_{rms} = \frac{I_0}{\sqrt{2}} \quad \text{and} \quad V_{rms} = \frac{V_0}{\sqrt{2}}$$

(See *mean power in a resistive load*.)

rolling is the motion of a body which moves across a surface without sliding. It involves a combination of translation and spin, such that the part of the body which, at any one time, is in contact with the surface is stationary. Rolling happens only on rough surfaces. On smooth surfaces, bodies slide.

root mean square speed is the square root of the *mean square speed*. (See also *kinetic theory of gases*, *distribution of speeds*.)

rotational energy is the kinetic energy associated with a rotating mass. It is calculated from the word equation:

rotational kinetic energy $= \frac{1}{2} \times$ moment of inertia $\times$ (angular speed)^2

$$W = \frac{1}{2} I \omega^2$$

rotational motion is the spinning motion of a body about a fixed axis. The motion is brought about by an unbalanced *torque* and the rate at which the angular speed increases is controlled by the *moment of inertia* of the spinning mass.

rubber is a naturally occurring polymer with molecules which are tangled and crooked. The molecules straighten when stretched and , once straightened, stop the rubber from stretching any further. If the tension is removed, the molecules curl up again and the rubber recovers its former shape. Natural cross-linking between the molecules within the rubber is effective but thinly scattered. The cross-linking is strong enough to enable the rubber to keep its shape in the unstressed condition and to stop it stretching still further once the molecules are straight. But the cross-links are so few that they offer no resistance to the straightening out of the molecules under even modest loads.

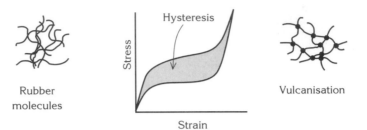

Rubber molecules

Vulcanisation

The loading curve on the load extension graph above differs from the unloading curve. The unloading curve shows a faster release of load with falling extension than would be expected. Since energy stored is area between the curve and the extension axis, less energy is released than is stored. The energy represented by the area enclosed by the loop remains in the specimen in the form of heat. This double curve is called a *hysteresis* loop. If you rapidly stretch and release a short section of rubber band and then test its temperature with your lips, it will be noticeably warmer than the rest of the band which was not stressed in this way. The capacity of a rubber specimen to withstand repeated elastic deformations without deteriorating is a measure of its *resilience*. (See *vulcanisation*.)

Do you know we also have A–Zs for:

- **Chemistry**
- **Biology**
- **Mathematics**
- **ICT & Computing?**

Ask for them in your local bookshop or see the inside back cover for ordering details.

sailing boats are driven by the flow of air over the sail. The diagram below shows a sailing boat seen from above. The wind flows faster over the outer, convex surface of the sail than over the inner, concave surface. According to *Bernouilli's principle*, this difference in speed sustains a difference in pressure, and the difference in pressure acting over the area of the sail gives a resultant thrust. Next, this thrust is resolved in two directions: along the keel and normal to the keel. The thrust normal to the keel is balanced by a normal contact force from the water. The thrust along the keel drives the boat through the water.

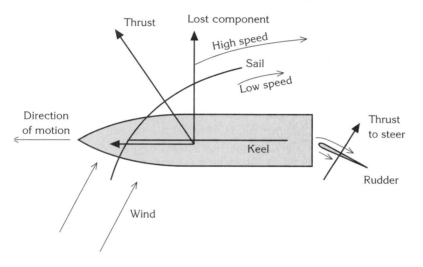

But the boat must be steered. This is achieved by the action of the water flowing over the two surfaces of the rudder, faster over the outside edge, provided that the rudder is held at an angle to the keel. The Bernouilli principle gives us a sideways thrust on the rudder which exerts a turning moment on the keel, and this in turn acts on to boat to change its direction of motion. This can be compared with *motor boats*.

sampling is the process of reading the instantaneous amplitude of a voltage signal successively at equal and very short intervals of time, usually by storing the instantaneous amplitude of the voltage on a small capacitor. Repetitive switching on and off of the contact between the voltage source and the capacitor creates a time sequence of discrete pulses which carry the information stored on the initial analogue signal from the voltage source. This is called pulse amplitude modulation (PAM). The sampled signal is discrete in time but analogue in amplitude. Sampling is also the first stage in any *analogue-to-digital conversion* process. See *Nyquist's theorem*.

satellite is a term which previously meant a planetary moon but now almost always means an artificial satellite in orbit around the Earth. Many of these are reconnaissance satellites in polar orbit between 700 km and 1000 km high with an orbital period of around 90 minutes and a useful life of up to ten years. A smaller number of satellites are geostationary satellites, in fixed positions above the equator.

Worked example

Calculate the height above the Earth of a geostationary satellite. The radius of the Earth is 6.37×10^6 m, its mass is 5.98×10^{24} kg and the universal gravitational constant, $G = 6.67 \times 10^{-11}$ N m^2 kg^{-2}.

The period of the satellite will be 24 hours since it encircles the Earth in the time the Earth takes to revolve once.

$$G\frac{mM}{r^2} = \frac{mv^2}{r} = \frac{m}{r} \times \left(\frac{2\pi r}{T}\right)^2$$

$$= \frac{m4\pi^2 r}{T^2}$$

$$r^3 = \frac{GMT^2}{4\pi^2}$$

so $\quad r = \sqrt[3]{\left(\frac{GMT^2}{4\pi^2}\right)} = 42.3 \times 10^6$ m

But this is the distance of the satellite from the Earth's centre. The distance above the Earth's surface is 35.9×10^6 m. Many numerical problems involving satellites can be worked out by starting with the first line in the worked example above.

(See *Newton's law of gravitation*.)

scalar quantity: a physical or mathematical quantity which has magnitude but no direction. Scalar quantities can be added algebraically. Examples include distance travelled, speed, density, pressure, energy, power, temperature, electric charge and electric potential. (See *algebraic sum, vector quantity*.)

scattering means a change in direction of a wave or particle without a significant amount of energy transferred to the scattering medium. An everyday example is the scattering of sunlight by the upper atmosphere to create a blue sky. It demonstrates above all that the scattering pattern is wavelength dependent. Red light goes mostly straight on. *Diffraction at a small circular hole* demonstrates that the preferred angles of the scattered beams (into a set of concentric rings) gives information about the scattering element.

Scattering, which contrasts with absorption in which the particle is annihilated, may be elastic (no energy loss) or inelastic (partial energy loss). The alpha particle scattering experiment provides an important example of elastic scattering. Low voltage (up to about 15 V) inelastic scattering of electrons by atoms in a gas was employed early in the twentieth century to verify the notions of excitation and ionisation of extranuclear electrons.

The best known scattering experiment, the alpha particle scattering experiment, is controlled by Coulomb repulsion. The *de Broglie wavelength* of alpha particles is too short compared with the nucleus to give a measurable diffraction effect. Electrons whose de

Broglie wavelengths were of the same order of magnitude as atomic diameters gave marked diffraction effects. The electron wavelengths achievable have been growing ever shorter with the advent of more and more powerful *particle accelerators*. The 200 MeV region is used for studying *nuclear radius* and nucleon diameter. 20 GeV electrons are used for what is called 'deep inelastic scattering'. These high energy electrons have wavelengths much smaller than the proton diameter and have been used to probe inside the proton. Using even higher voltages, quarks have been shown to be less than 10^{-18} m in size. Protons and other particles are used in scattering experiments too.

Where the scattering is inelastic, some kind of interaction and energy transfer is taking place and physicists talk about collision experiments. The emphasis in collision experiments is on learning how particles interact and on the discovery of new particles. See *electron diffraction*.

second, symbol s, is a *base unit* in *SI*. It is the SI unit of time. It equals 9 192 631 770 periods of the radiation corresponding to the transition between the two hyperfine ground state levels of the caesium–133 atom.

second law of thermodynamics: a principle which underpins the theory of heat engines. One statement of the law is the following: an engine cannot produce mechanical work with no effect other than the withdrawal of heat from the coolest body in the surroundings. Two important consequences of the law are the definition of the *absolute thermodynamic temperature scale* and the well known expression $[T_1 - T_2]/T_1$, for the efficiency of a perfect heat engine. The law is not tested at A level. See *heat engine concept*.

secondary fuels are fuels such as electricity or petrol which draw their energy from *primary fuels* but provide the consumer with a more convenient form of energy. The production of secondary fuels from primary fuels always involves an energy loss. (See also *energy sources*.)

self-inductance is a measure of the e.m.f. induced in a circuit by a *change of flux linkage* brought about by a change of current in the same circuit. The symbol for self-inductance is L and the SI unit is the henry (H).

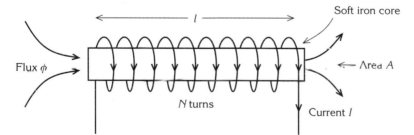

The diagram above shows a coil with N turns of length l and cross-sectional area A, wound on a soft iron core. The current I in the coil sets up a magnetic flux ϕ in the core and a flux linkage of $N\phi$. If the current I changes, the flux linkages $N\phi$ will also change. There will be set up within the coil a back e.m.f. E the magnitude of which is given by the laws of *electromagnetic induction*:

$$E = -N\frac{d\phi}{dt}$$

But

$$\phi = BA = \frac{\mu_0 NI}{l}.A$$

where μ_0 is the permeability of free space, so

$$E = -\frac{\mu_0 N^2 A}{l}.\frac{dI}{dt}$$

Next, we define the coefficient of self-inductance L of the coil by the equation:

$$E = -L.\frac{dI}{dt}$$

We see that

$$N\phi = LI$$

and that

$$L = \frac{\mu_0 N^2 A}{l}$$

The coefficient of self-inductance L of a coil can be defined either as the induced e.m.f. per unit rate of change of current or as the flux linkages in the coil per unit current. These alternatives can be written out in the form of word equations:

$$\text{coefficient of self inductance } L = \frac{\text{e.m.f. induced in the coil}}{\text{rate of change of current}}$$

$$= \frac{\text{flux linkage for coil}}{\text{current in coil}}$$

The first of these word equations can be used to define the henry (H), the unit of self-inductance:

$$\text{henry} = \frac{\text{volt}}{\text{ampere per second}} = \text{ohm second}$$

That is, the henry (H) is the same as the ohm second.

It is worth going back to the equation $L = \mu_0 N^2 A/l$ and looking at the units. N is a number and has no unit. So the unit for μ_0 is the same as the unit for LI/A, i.e. H m^{-1}.

(See *mutual inductance*, *inductive reactance*, *series resistance and inductance*, *resonance in AC circuits*.)

semiconductor: a material which is an insulator at absolute zero but which has valence electrons (the outer electrons) loosely enough bound for a small proportion to be released by thermal motions at room temperature. Such materials are called intrinsic semiconductors. Each electron freed in this way leaves a positive mobile 'hole' in the lattice. Electrons and holes both carry current (though electrons have the higher mobility so they carry more current) and are present in equal numbers.

The semiconducting properties of an intrinsic semiconductor can be drastically changed by the introduction of small quantities of impurity, a process sometimes known as doping.

These impurities bring about either a sharp increase in the number of conduction electrons (*n*-type) or a sharp increase in the number of holes (*p*–type). The resulting material is called an extrinsic conductor.

A major use of semiconducting materials is linked to the electrical behaviour at the junction between two semiconducting materials, one *n*-type and the other *p*-type, sealed together. The properties of the so-called *p*–*n* junction lie at the root of the semiconductor revolution. (See *semiconductor diodes*.)

semiconductor diodes are formed by fusing *n*-type and *p*-type semiconducting materials to form a series pair of conductors. Electrons flow across the junction in one direction but not in the other so the diode conducts current in one direction only.

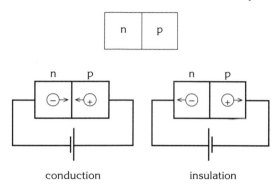

conduction   insulation

If the *n*-type material is negative with respect to the *p*-type, electrons flow across the junction from *n* to *p*, holes flow the other way and the diode conducts. For silicon-based materials, the negative bias of the *n*-type material must exceed about 0.6 V before the current climbs to an appreciable level. It then increases strongly with increasing bias. This is illustrated by the following typical current–voltage characteristic.

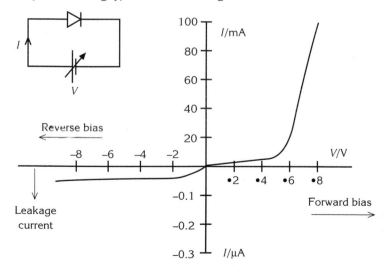

If the *n*-type material is positive with respect to the *p*-type, electrons in the *n*-type and holes in the *p*-type move away from the junction. Nothing flows across the junction (apart from a residual current of a few microamps, called leakage current) and a potential difference is set up across the junction which opposes the applied potential difference. If too large a reverse voltage is applied to the diode it breaks down. Some diodes, called Zener diodes, exploit this effect. (See *semiconductors*.)

sensitivity is a measure of the smallest change in a physical quantity that a measuring instrument can detect. This is a separate matter from reliability. A mercury thermometer, for instance, will indicate a change in temperature as small as a tenth of a degree kelvin, while its reading may differ from the true value by more than a full degree. *Precision* is an expression of the inherent uncertainty in an instrument and takes into account both the sensitivity and the overall reliability. A good instrument is accompanied by a written definition of the precision guaranteed by the manufacturer.

The word sensitivity is used also as a closely defined technical expression of the scale size for some instruments, particularly ammeters and voltmeters. Sensitivity is defined as the deflection (in radians) per ampere or per volt or, alternatively, in scale divisions per ampere or per volt. The ratio is called current sensitivity or voltage sensitivity. The numerical value varies enormously with the range of the instrument (is it calibrated in, for example, amperes or microamperes?) and needs careful interpretation.

series resistance and capacitance circuits take the current I in the circuit, which is common to all series components, as the reference. The current is assumed to be of angular frequency ω and to vary with time in the form of a sine wave:

$$I = I_0 \sin \omega t$$

The circuit shows a capacitor, capacitance C, in series with a resistor, resistance R, and an AC power supply.

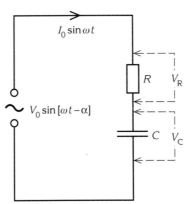

The voltage ($VR \sin \omega t$) across the resistor is in phase with the current. It is given by Ohm's law:

$$V_R \sin \omega t = IR = I_0 R \sin \omega t$$

and the voltage ($-V_C \cos \omega t$) across the capacitor, which lags the current by a quarter of a cycle, is written:

$$-V_C \cos \omega t = -\frac{I_0}{\omega C} \cos \omega t$$

To find the supply voltage from V_R and V_C, a phasor diagram is drawn (see below). The lengths of the phasors equal the peak values of the different voltages.

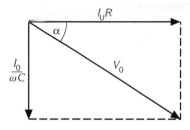

The supply voltage V_0 is found with Pythagoras' theorem:

$$V = V_0 \sin (\omega t - \alpha)$$

$$V_0 = \sqrt{\left\{ (I_0 R)^2 + \left(\frac{I_0}{\omega C} \right)^2 \right\}} = I_0 \sqrt{\left\{ R^2 + \frac{1}{\omega^2 C^2} \right\}}$$

$$\tan \alpha = \frac{V_C}{V_R} = \frac{1}{\omega CR}$$

$$Z = \sqrt{\left\{ R^2 + \frac{1}{\omega^2 C^2} \right\}} = \frac{V_0}{I_0}$$

where α is the phase lead of the current over the supply voltage and Z, the ratio of peak voltage to peak current, is the impedance of the circuit.

Many problems can be solved by calculating the impedance and then working out the peak current from the peak voltage. Divide the peak values by $\sqrt{2}$ if you need the r.m.s. values. (See *capacitive reactance*, *inductive reactance*, *series resistance and inductance*, *resonance in AC circuits*.)

series resistance and inductance circuits take the current I in the circuit, which is common to all series components, as the reference. The current is assumed to be of one angular frequency ω and to vary with time in the form of a sine wave:

$$I = I_0 \sin \omega t$$

The circuit below shows an inductor, inductance L, in series with a resistor, resistance R.

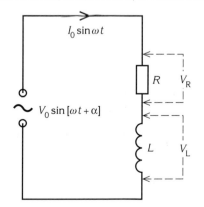

The voltage (V_R sin ωt) across the resistor is in phase with the current. It is given by Ohm's law:

$$V_R \sin \omega t = IR = I_0 R \sin \omega t$$

and the voltage (V_L cos ωt) across the inductor, which opposes the back e.m.f. in the inductor and which leads the current by a quarter of a cycle, is derived as follows:

$$-(-L \frac{dI}{dt}) = \omega L I_0 \cos \omega t = V_L \cos \omega t$$

To find the supply voltage from V_R and V_L, a phasor diagram is drawn (see below). The lengths of the phasors equal the peak values of the different voltages.

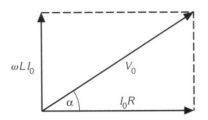

The supply voltage V_0 is found with Pythagoras' theorem:

$$V = V_0 \sin (\omega t + \alpha)$$

$$V_0 = \sqrt{\{(I_0 R)^2 + (I_0 \omega L)^2\}} = I_0 \sqrt{\{R^2 + \omega^2 L^2\}}$$

$$\tan \alpha = \frac{V_L}{V_R} = \frac{\omega L}{R}$$

$$Z = \sqrt{\{R^2 + \omega^2 L^2\}} = \frac{V_0}{I_0}$$

where α is the phase lead of the supply voltage over the current and Z, the ratio of peak voltage to peak current, is the impedance of the circuit.

Many problems can be solved by calculating the impedance and then working out the peak current from the peak voltage. Divide the peak values by $\sqrt{2}$ if you need the r.m.s. values. (See *capacitive reactance, inductive reactance, series resistance and inductance, resonance in AC circuits*.)

SI is the agreed abbreviation for Système International d'Unités. This system provides the international legal framework for the measurement of physical quantities and for the units used to define amounts of physical quantities. (See *base quantity, base unit, derived quantity, derived unit*.)

simple harmonic motion is the motion of a body which is acted upon by a resultant force whose magnitude is proportional to the distance of the body from a fixed point and whose direction is always towards that point.

This statement can be put in algebraic form:

$$F = -kx$$

where F is the restoring force on the body, x is the displacement of the body from its rest position and k is the restoring force per unit displacement. The unit for k is N m^{-1}.

The algebra is simpler if the equation is written in the form:

$$F = -m\omega^2 x = -m(2\pi f)^2.x$$

where m is the mass of the body on which the force F acts, f is the frequency of the motion and ω is the angular frequency of the motion. If a is the acceleration when displacement is x, then:

$$a = -\omega^2 x = -(2\pi f)^2.x$$

If this equation is solved with time measured from an instant when the mass is at maximum displacement (i.e. $x = x_0$ when $t = 0$) in the positive x direction, then we get the following important results:

$$x = x_0 \cos \omega t \quad = x_0 \cos 2\pi f t$$

$$v = -\omega x_0 \sin \omega t \quad = -2\pi f x_0 \sin 2\pi f t \quad = -2\pi f \sqrt{(x_0^2 - x^2)}$$

$$a = -\omega^2 x \quad = -(2\pi f)^2 x_0 \cos 2\pi f t$$

$$T = \frac{1}{f} \quad = \frac{2\pi}{\omega}$$

$$E_k = \frac{1}{2}mv^2 \quad = \frac{1}{2}m(2\pi f)^2(x_0^2 - x^2)$$

where x_0 is the maximum value of the displacement (i.e. the amplitude of the motion) and T is the period. Notice that the peak value of the kinetic energy is proportional to the square of the amplitude.

The diagram on page 224 shows the different graphs which come from these equations and which you should know about.

The secret of solving problems on simple harmonic motion is to concentrate on ω. If you are given the period T, then find ω and use it as part of the data in the other equations to calculate whatever the question asks. If you are required to find T, use the data and the earlier equations to find ω and then T.

Worked examples

1. An object which is oscillating with simple harmonic motion has period 2.2 seconds and amplitude 5.1 cm. Assuming that the displacement is maximum at time $t = 0$, find the displacement and the speed at time $t = 0.80$ s.

   Using $T = 2.2$ s, we have $\omega = 2\pi/T = 2.86$ rad s^{-1}.

   At $t = 0.80$ s, $\omega t = 2.29$ radians or $131°$, $\cos \omega t = -0.657$, and

   displacement $x = -3.35$ cm.

   Using this value for x in the velocity formula, we have
   velocity $v = -11.0$ cm s^{-1}.

   We can now state the result in SI units and to the correct number of significant figures:

   displacement at time $t = 0.80$ s is 0.034 m

   and the velocity is -0.110 m s^{-1}.

2. An object which is oscillating with simple harmonic motion of amplitude 0.122 m has speed 0.38 m s^{-1} when the displacement is 0.076 m. Calculate the period of the motion.

Using the velocity formula, we have:

$$v = \omega\sqrt{(x_0^2 - x^2)}$$

$$\omega = \frac{0.38 \text{ m s}^{-1}}{\sqrt{[(0.122 \text{ m})^2 - (0.076 \text{ m})^2]}} = 3.98 \text{ rad s}^{-1}$$

The period can now be found:

$$T = \frac{2\pi \text{ rad}}{3.98 \text{ rad s}^{-1}} = 1.58 \text{ s} = 1.6 \text{ s} \quad \text{(to 2 s.f.)}$$

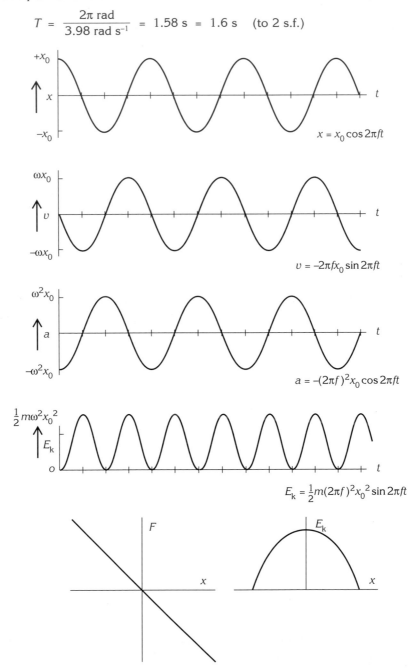

$x = x_0 \cos 2\pi ft$

$v = -2\pi f x_0 \sin 2\pi ft$

$a = -(2\pi f)^2 x_0 \cos 2\pi ft$

$E_k = \frac{1}{2}m(2\pi f)^2 x_0^2 \sin 2\pi ft$

(See *angular frequency, frequency, simple pendulum, helical spring, damped harmonic motion, resonance*.)

simple pendulum: a small dense object, usually a metal sphere, suspended from a fixed point by a thread and free to move from side to side. If T is the period of the motion, l is the distance from the point of support to the centre of mass of the object and g is the acceleration of free fall, then:

$$T = 2\pi\sqrt{\left(\frac{l}{g}\right)}$$

Notice that T is independent of the amplitude. This is an approximate result only but it is accurate enough for angles of swing smaller than about 5°. Notice too that the definition of the length l is not just the length of the supporting thread; it is a particular distance and this should be labelled precisely on any diagram.

Experiment: to measure the acceleration of free fall using a simple pendulum

Apparatus

The diagram should show the simple pendulum suspended from a rigid support. The length l, that is the distance from the rigid support to the centre of mass of the pendulum bob, must be clearly defined in the diagram and labelled. There should be some kind of visible indicator at the rest position to allow timings to start and stop as the pendulum bob goes through the centre of its swing.

Measurements

For each of a range of lengths, measure the distance l with a metre rule and the period T. The distance l should be measured carefully three times and an average taken. The period T should be found by measuring the time needed for 25 complete swings three times for each length and then dividing the average of these three measurements by 25.

Conclusion

Plot a graph of T^2 in second^2 against l in metres. It will be a straight line of gradient $4\pi^2/g$ which passes through the origin. The acceleration of free fall g is $4\pi^2$ divided by the gradient.

Comment

The chief source of error is the random error in the starting and stopping of the clock. This is reduced by taking the average of three measurements and by counting the periods as the pendulum bob passes through the rest position.

A careful student should be able to reach the point at which the uncertainties linked to the measurements are outweighed by the fresh uncertainties introduced by using an ordinary sized sheet of graph paper.

(See *simple harmonic motion*.)

sizes of atoms are all in the region of $n \times 10^{-10}$ m where n is a small number. (See *oil film experiment*.)

slip: a permanent dislocation in the crystal structure of a metal. It happens during the plastic flow stage for a metal stressed beyond its elastic limit. It takes the form of adjacent crystal planes sliding over one another in the presence of excessive stress. Slip lines can be seen with a microscope on the surfaces of many metal wires which have been stretched beyond their elastic limits. The extreme case is for a wire of a metal such as cadmium which can be prepared as a single crystal. The forces within the crystal planes are stronger than the forces holding the planes together. Planes where sliding occurs are called slip planes. See *plastic deformation, cleavage*.

smoothing capacitor: see *rectification, half-wave and full-wave*.

Snell's law: see *refraction of light*.

solar constant is the incident power delivered by radiation from the Sun per square metre at right angles to the direction of flow at the top of the atmosphere near the equator. The accurate value is under review; a working figure is 1.37 kW m^{-2}.

It is a difficult measurement to make. The 'top of the atmosphere' needs a closer definition. The distance of the Earth from the Sun varies during the year; the Sun is given to occasional and erratic changes.

solar panels are devices for absorbing solar radiation across an appreciable area and storing it thermally in water which circulates through the panels.

solar power is power drawn directly from the sun as in *solar panels* used for home heating or solar cells used to produce electricity directly from solar radiation. (See *energy sources, photoelectric cells*.)

solenoid: a tight helix of conducting wire. There may be one or more layers of turns. Provided that the length of the solenoid is more than ten times its diameter, a uniform magnetic field runs along the inside of the solenoid of strength B, where:

$$B = \mu_0 n I = \frac{\mu_0 N I}{l}$$

μ_0 is the permeability of vacuum, I is the current, n is the number of turns per metre length of solenoid, N is the number of turns and l is the length of the solenoid.

The form of the magnetic field set up in a solenoid is shown in the diagram below. The direction of the *lines of force* within the solenoid is given by the *right hand grip rule*. Grip the solenoid with your right hand and with your fingers pointing along the direction of the current. Your thumb runs along the solenoid in the same direction as the magnetic field lines *inside* the solenoid.

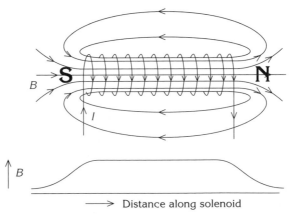

Remember, when copying this diagram, to include at least five equally-spaced field lines and to leave the breaks in the coils to show the direction of the windings. (Omitting either of these points could cost you all the diagram marks.) The graph in the diagram reminds us that the magnitude of the magnetic field strength begins to tail off just before the end of the solenoid is reached.

The magnetic field strength within a solenoid can be increased by filling the space within the solenoid with a *ferromagnetic* material. The magnetic field strength within the solenoid becomes:

$$B = \mu_r \mu_0 nI = \frac{\mu_r \mu_0 NI}{l}$$

where μ_r is called the *relative permeability* of the ferromagnetic material.

This may not seem of much use if the new, stronger magnetic field cannot be accessed because of the ferromagnetic core. It is however, useful in two ways. Both the magnetic field outside the solenoid and the *self-inductance* of the solenoid are increased in the same proportion. Electromagnets and tuning coils are solenoids with ferromagnetic cores. (See *permeability of free space, relative permeability*.)

sound waves are sets of pressure waves which pass through the air, one after the other, at a speed governed by the density of the air and its elasticity. The first diagram below illustrates the essential events leading up to the formation of a single pressure pulse. A loudspeaker cone moves sharply to the right. This affects the air on the right of the loudspeaker. A compression is formed (a localised rise in density and pressure) and the air in the compression has momentum towards the right and potential energy associated with the higher pressure. The momentum equals the impulse imposed on the air by the loudspeaker cone and it will carry the compression away from the cone as the momentum is passed from one layer of air to the next. There is a flow of sound energy and of momentum. There is no net displacement of air.

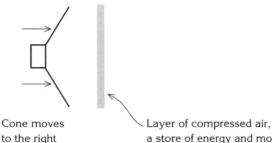

Cone moves
to the right

Layer of compressed air,
a store of energy and momentum

The next step is to consider what happens if the loudspeaker moves repetitively backwards and forwards. Assume that the loudspeaker moves with simple harmonic motion along the axis of its cone. Each time the cone moves to the right it sets up a new compression pulse which follows its predecessors into the air space and away from the cone. The speed of these compressions is fixed by the elasticity and density of the air. The speed of sound in air is a property of air, nothing else. A more vigorous motion from the loudspeaker gives a tighter compression with a higher pressure at its centre; the compression will not move through the air any faster but it will carry more energy.

The air particles in the path of the sound wave oscillate backwards and forwards along the line of energy flow. Such waves are called *longitudinal waves*. There is no net motion of the air arising from the passage of the sound wave. The oscillations of the air particles are an ordered motion superimposed upon the disordered motion dealt with in *kinetic theory of gases*. The two motions are independent of one another.

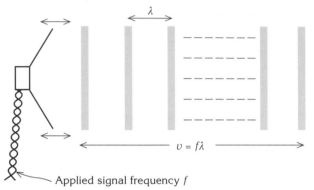

Applied signal frequency f

The number of compressions issuing from the loudspeaker per second equals the frequency of the motion of the loudspeaker (and this, in turn, equals the frequency of the electrical signal applied to its coil). The distance moved by one compression in a second (the speed) divided by the number of compressions created per second equals the average distance between the centres of successive compressions. This is called the wavelength of the sound. It follows that:

speed of sound = frequency × wavelength

$$c = f \times \lambda$$

This relation is true for any kind of wave. Speed depends on the medium, frequency depends on the source and wavelength is decided by speed and frequency together. The four graphs in the following figures split into two pairs.

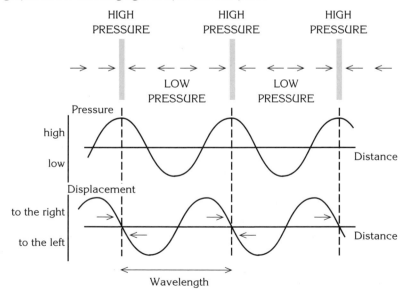

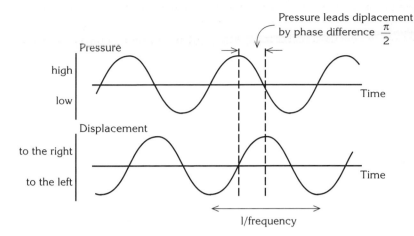

Pressure leads diplacement by phase difference $\frac{\pi}{2}$

The first pair shows how the pressure and the displacement vary with distance along the path of the sound wave at a particular instant. The second pair shows how the values of the same two variables, pressure and displacement, vary with time at a particular spot in the path of the wave. The first pair of graphs illustrates the meaning of wavelength, the second pair illustrates period and phase. None of the graphs has start or finish points; the sound wave is a continuum.

The pressure amplitude in a sound wave varies between about 50 Pa for a loud sound (atmospheric pressure is about 100 000 Pa or two thousand times higher) and 10^{-5} Pa for a very weak but still audible sound. Loudness is not expressed as a pressure change, but in terms of the power transmitted by the sound wave per unit area. The power transmitted ranges from about 1 W m^{-2} for a loud sound down to 10^{-12} W m^{-2} for a barely audible sound. (See *loudness*.)

special relativity brings together the physics of motion, optics and electromagnetism into a single scheme of things. The results cannot be applied to systems which are accelerating relative to one another. This is why a theory of *general relativity* is needed.

Imagine an observer called Jack in a laboratory equipped with good quality measuring instruments. The laboratory appears stationary to Jack: it is his frame of reference for data collection and study.

A second observer, Jill, is in an identical laboratory. This laboratory, Jill's reference frame, is travelling with constant velocity v relative to Jack's reference frame, his laboratory.

Both observers are equipped with identical helical springs of period of oscillation Δt, and with identical measuring rods of length Δl. The measuring rods are mounted parallel to the velocity vector v of Jill's frame of reference as seen by Jack.

Jack checks the period of his own spring and the length of his own measuring rod, and finds them unchanged. Jill finds the same.

However, when Jack uses his instruments to check Jill's helical spring, he finds that its period has lengthened to $\Delta t'$. So, clocks in Jill's reference frame seem to be running slow when viewed by Jack from his reference frame. Similarly, Jill's measuring rod appears to have shortened compared with Jack's own measuring rod to a new length $\Delta l'$. Jill's view of Jack's helical spring and of his measuring rod yield exactly the same changes.

This situation is analysed on the basis of two new postulates:

For any two observers whose reference frames are in a state of uniform motion relative to one another (i.e. for any two inertial frames of reference):

- the laws controlling mechanical, optical and electromagnetic events must be the same
- the speed of light in vacuum and in all directions is the same.

Two important relationships can be derived on the basis of these two postulates:

$$\Delta t' = \frac{\Delta t}{\sqrt{\left(1 - \dfrac{v^2}{c^2}\right)}}$$

$$\Delta l' = \Delta l \times \sqrt{\left(1 - \frac{v^2}{c^2}\right)}$$

Jack and Jill both record values Δt and Δl when using their own instruments to measure a given time interval and a given length in their own reference frames. They both record values $\Delta t'$ and $\Delta l'$ when measuring the same time interval and the same distance in the other's reference frame. The velocity of one reference frame relative to the other is v, Δl is parallel to v in both reference frames and c is the speed of light.

Another important result of special relativity is mass–energy equivalence as summarised by the equation

$$\Delta E = \Delta mc^2$$

This equation tells us that if we warm up a football (supply energy ΔE joules), then the football will be harder to kick a given distance because its inertial mass has increased by an amount Δm kg as calculated from the mass–energy equivalence equation. A non-trivial example is the fission of a uranium-235 atom following a collision with a slow neutron:

$$^{235}_{92}U + ^{1}_{0}n \longrightarrow ^{236}_{92}U \longrightarrow ^{140}_{54}Xe + ^{94}_{38}Sr + 2\ ^{1}_{0}n + \gamma + 200 \text{ MeV}$$

Mass–energy is conserved throughout this change. But the rest mass of the products is less than the rest mass of the reactants. There has been a transformation. The energy released equals Δmc^2 and until that energy is lost from the system, the total mass of the energetic particles is the same as the mass of the reactants. By total mass we mean the mass of the system as decided by the force needed to induce a given acceleration. Equally, the recipient of that lost energy, once it escapes from the system, experiences a corresponding mass increase.

specific heat capacity is the heat needed to raise the temperature of one kilogram of a material by one degree kelvin. It is a physical property of a material. The SI unit is J kg^{-1} K^{-1}. The symbol used is c (note the lower case form). The word equation definition is:

$$\text{specific heat capacity} = \frac{\text{heat supplied}}{\text{mass} \times \text{temperature rise}}$$

$$\text{unit for specific heat capacity} = \frac{\text{joules}}{\text{kilograms} \times \text{kelvin}} \quad \text{or J kg}^{-1} \text{ K}^{-1}$$

$$C = \frac{\Delta Q}{m \times \Delta\theta}$$

Worked examples

1. Calculate the heat needed to raise the temperature of a 0.250 kg piece of copper (specific heat capacity of copper = 380 J kg^{-1} K^{-1}) from 12°C to 47°C.

   $\Delta Q = mc\Delta\theta = (0.250 \text{ kg})(380 \text{ J kg}^{-1} \text{ K}^{-1})(35 \text{ K}) = 3.3 \text{ kJ}$

   (By putting the answer in kJ, the problem of too many significant figures is neatly avoided.)

2. An aluminium vessel of mass 0.035 kg contains 0.052 kg of water at 16°C. A 0.050 kg block of aluminium at 100°C is dropped into the water. The specific heat capacities of aluminium and water are 900 J kg^{-1} K^{-1} and 4200 J kg^{-1} K^{-1} respectively. Calculate the final temperature of the water.

   Let θ represent the final temperature of the water and the aluminium:

$$\begin{array}{c} \text{heat lost} \\ \text{by aluminium block} \end{array} = \text{mass} \times \begin{array}{c} \text{specific heat} \\ \text{capacity} \end{array} \times \begin{array}{c} \text{temperature} \\ \text{drop} \end{array}$$

$$= (0.050 \text{ kg}) \times (900 \text{ J kg}^{-1} \text{ K}^{-1}) \times [(100 - \theta) \text{ K}]$$

$$= (4500 - 45\theta) \text{ J}$$

$$\begin{array}{c} \text{heat gained by aluminium} \\ \text{vessel and water} \end{array} = \begin{array}{c} [(0.035 \text{ kg})(900 \text{ J kg}^{-1} \text{ K}^{-1}) + (0.052 \text{ kg}) \\ \times (4200 \text{ J kg}^{-1} \text{ K}^{-1})] \times (\theta - 16) \text{ K} \end{array}$$

$$= (250\theta - 4000) \text{ J}$$

$$(4500 - 45\theta) = (250\theta - 4000)$$

$$295\theta = 8500$$

$$\theta = 29°C$$

The principle of the second example is to assume that once the aluminium block is dropped in the water, the internal energy of the whole system is constant. The internal energy lost by the block equals that gained by the vessel and the water. This will be true only if the lagging of the apparatus is perfect. (See also *specific heat capacity (measurement of), heat capacity.*)

specific heat capacity (measurement of): these experiments need only be known in general outline. The two principal kinds of experiment are a 'method of mixtures' and a method based on electrical heating.

The method of mixtures builds on the assumption that the specific heat capacity of water and possibly the heat capacity of a metal vessel (called a calorimeter) are known already. A known mass of water is poured into the calorimeter; its temperature is measured with a mercury thermometer. The mass of the solid or liquid body the specific heat capacity of which is needed, is found by weighing. The body under examination is then heated to 100°C by immersion in steam or boiling water. It is quickly transferred to the calorimeter and the maximum temperature of the mixture recorded. The gain in internal energy of the water and calorimeter is calculated. This result is divided by both the mass of the body and its drop in temperature to find the specific heat capacity. The principal difficulty is to ensure that there is no serious heat loss to the surroundings while the temperature settles down to its maximum value. This is achieved by having excellent lagging, a lid to discourage evaporation and a sensible choice of liquid and solid masses.

In the electrical method, the temperature of a known mass of material is raised by heating it electrically for a known period of time. The rise in temperature is measured with a mercury thermometer. The thermal energy transferred to the mass is IVt, the product of the current I in the heater, the voltage V across the heater and the time t of current flow. Divide the thermal energy transferred, IVt, by both the mass m of the body and the temperature rise θ and you are left with the specific heat capacity c of the body. For a liquid, the experiment is repeated twice using the same vessel but different amounts of liquid. Extra electrical energy is needed to heat the increased mass of liquid by the same amount as before but the container accounts only for the same amount as before. This extra electrical energy divided by the extra mass and the temperature rise gives the specific heat capacity.

specific latent heat is the heat transferred per unit mass when a pure material changes state without changing temperature. The SI unit of specific latent heat is joules per kilogram (J kg⁻¹). The symbol used is L. The specific latent heat of fusion of ice (or of freezing of water) is 334 kJ kg⁻¹ and the specific latent heat of vaporisation of water (or of condensation of steam) is 2.26 MJ kg⁻¹. The word equation definition is:

$$\text{specific latent heat} = \frac{\text{heat transferred}}{\text{mass}}$$

$$\text{unit} = \frac{\text{joules}}{\text{kilograms}} = \text{J kg}^{-1}$$

Worked example

Calculate the thermal energy transferred when 350 g of ice at 0°C is transformed into steam at 100°C.

$$\text{heat transferred} = mL_{ice} + mc\theta + mL_{steam}$$
$$= (0.350 \text{ kg})(334\,000 \text{ J} + 420\,000 \text{ J} + 2\,260\,000 \text{ J})$$
$$= 1.05 \text{ MJ}$$

1.05 MJ is enough heat to raise the temperature of 2.5 kg of water from the freezing point to the boiling point without change of state. In the above calculation, only a seventh of the heat supplied was used to change the temperature; six sevenths were used to change the state. The large amount of heat stored as latent heat in the water vapour in the atmosphere keeps the Earth's climate manageable by transferring energy from the equator to the poles. Were the specific latent heat of vaporisation appreciably smaller, temperature differences from pole to equator would be greater and storm violence would increase dramatically.

speed is a *scalar quantity*. It is defined by the word equation:

$$\text{speed} = \frac{\text{distance travelled}}{\text{time taken}}$$

Distance travelled must be measured along the travel path and cannot decrease. A step backwards is another step: it adds to the total distance travelled. Since distance travelled is always a positive quantity, so is speed. The word equation that defines speed also defines its unit:

$$\text{SI unit for speed} = \frac{\text{unit for distance travelled}}{\text{unit for time}} = \frac{\text{metre}}{\text{second}} = \text{m s}^{-1}$$

Uniform speed means equal distances travelled in equal times.

Average speed is defined by the word equation:

$$\text{average speed} = \frac{\text{total distance travelled}}{\text{total time taken}}$$

Instantaneous speed is the average speed over a very small time interval at a given instant. It equals the gradient of the tangent to the *distance–time graph* at that instant.

A graph of instantaneous speed against time is called a speed–time graph. The area between a speed–time curve and the time axis equals the distance travelled between the corresponding times. The gradient of the tangent to a *speed–time graph* equals the rate of change of speed at that instant.

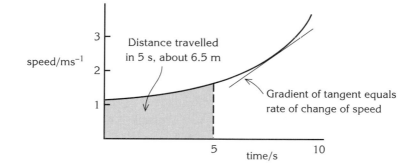

Rate of change of speed is a scalar quantity with units m s⁻². Rate of change of speed is not acceleration. Acceleration is a *vector quantity*; it is rate of change of velocity. A body moving along a circular path at constant speed has acceleration even though the rate of change of speed is zero.

Worked example

A car, starting from rest at point A, increases its speed from zero to 12.5 m s⁻¹ at a uniform rate in 12 seconds. It travels at this speed of 12.5 m s⁻¹ for a further 20 seconds. Finally, it slows down uniformly to rest at B in 20 seconds.

Look at the speed–time graph on page 234. The line for the first 12 seconds is straight and slopes upwards. The gradient is constant and positive. The speed is increasing at a constant rate. For the next 20 seconds the speed is constant. The gradient of the speed–time curve is zero. For the final 20 second segment, the gradient is constant but negative; the car is slowing down. Speed cannot be negative but its rate of change can.

The area to the left of the upward line at 15 seconds equals the distance travelled by the car in the first 15 seconds. Imagine this line travelling slowly along the time axis from 0 seconds to 52 seconds. The growing area to its left represents the distance travelled and is plotted on the lower graph. This area can only increase. Distance once travelled cannot be untravelled.

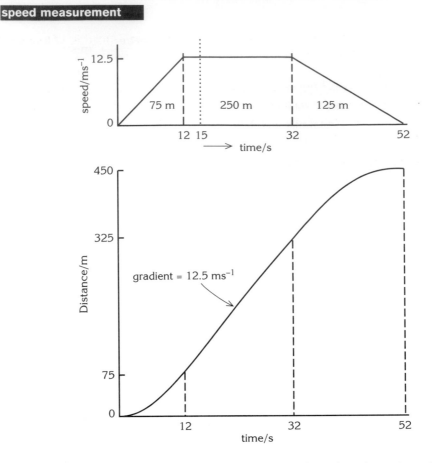

The distance travelled, found in this way, can be plotted against time. The distance–time graph so found, cannot have a negative gradient because speed cannot have a negative value. The curved ends of the line represent increasing speed at the beginning and decreasing speed at the end.

From 12 seconds to 32 seconds, the line is straight. The constant gradient corresponds to a constant speed of 12.5 m s⁻¹.

speed measurement involves measuring a distance and the corresponding time interval. When describing how speed is measured in an experiment:

● define the distance

● state how it is measured (usually with a metre rule)

● state how the time interval is recognised (usually by the breaking of a light beam)

● and how it is measured (usually by a digital timer running while the light beam is broken).

If you are using the ticker-tape method, the same general principles apply but the details are different.

speed of a wave: the speed at which energy is transmitted by the wave. (See *speed of sound*, *speed of light*.)

speed of light is, by definition, 299 792 458 m s⁻¹ exactly. (See *permittivity of free space, permeability of free space, the ampere.*)

speed of sound: the distance sound waves travel per unit time. The phrase 'speed of sound in free air' implies the speed at which sound travels across an open laboratory or an open space. Examination questions asking for a description of a method of measuring the speed of sound in free air rarely allow any marks for answers based on a resonance tube method or for answers of the 'handkerchief and starting pistol' variety.

Experiment: to measure the speed of sound

Apparatus

A suitable arrangement is shown in the diagram.

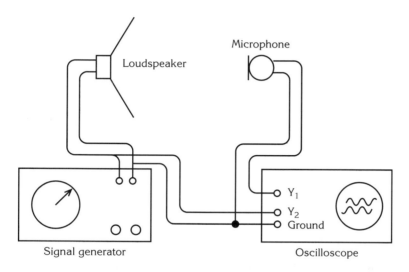

The output from the signal generator is set at about 1 kHz. It is connected to the Y2 input of a double beam oscilloscope and to the loudspeaker. The time base is adjusted to show about three full cycles of the signal from the signal generator. The microphone is connected to the Y1 input.

Measurements

Place the microphone, facing the loudspeaker, at a distance for which the two wave traces on the oscilloscope screen are in phase. Mark the position of the microphone. Move the microphone away from the loudspeaker slowly and note how the signals get increasingly out of phase until they are back in phase again. Continue moving the microphone until this has happened four times. Measure the total distance moved by the microphone with a metre rule. This equals four wavelengths. Repeat this measurement three or four times and find an average value for the wavelength.

A simple method for measuring the frequency of the signal generator output is shown in the diagram on page 236.

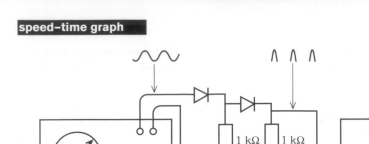

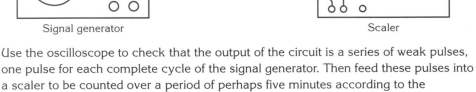

Signal generator   Scaler

Use the oscilloscope to check that the output of the circuit is a series of weak pulses, one pulse for each complete cycle of the signal generator. Then feed these pulses into a scaler to be counted over a period of perhaps five minutes according to the capacity of the counter. Dividing the final count by 300 seconds gives the frequency.

Conclusion

The product of the wavelength and the frequency equals the speed of sound.

Comment

It would be a simple matter to repeat the experiment with different frequency settings. The measurements could be used to check that the speed of sound is the same for all frequencies.

---

speed–time graph: a graphical display of the time variation of *speed*. The area between the speed–time curve and the time axis gives distance travelled. The gradient of the tangent to the curve gives the rate of change of speed with time. This is not the same thing as *acceleration*; acceleration is a *vector quantity*.

It sometimes helps to remember that speed is what a speedometer in a car records. It tells you how fast a car is going but not its direction of travel. The magnitude of speed can rise or fall but it cannot have a negative value. A speed–time graph cannot cut the time axis.

(See *distance, displacement, velocity–time graphs*.)

spherical aberration is a distortion of an image formed by a lens or mirror. It originates in the smaller focal lengths at the edges of the lens or mirror compared with the focal length of the part near the principal axis. The fault can be corrected by a small change in the shape of the lens, but this is costly. The reduction of spherical aberration is just one more reason why higher quality telescopes or binoculars are expensive.

spontaneous is a word that is often used in connection with radioactive decay. The word implies that the decay of a particular nucleus is instantaneous, there is no warning and there seems to be no external trigger. It is linked to the idea of randomness. Radioactive decay is controlled by strong or weak interactions and these have short ranges with influences that do not extend outside the nucleus or beyond isolated nuclear particles.

spring constant, k, is defined by the equation $F = kx$ where F is the tensile force applied to a spring and x is its extension. The unit is N m^{-1}. (If you study mechanics in A level mathematics, take care not to confuse spring constant k with spring modulus λ defined by the equation $F = \lambda x/L$ where L is the relaxed length of the spring.) See *helical spring*.

standard pressure is the bar which is defined as exactly 10^5 Pa. The quantity, $1.013\,25 \times 10^5$ Pa exactly, is called the standard atmosphere. (See also *pascal, bar, standard temperature and pressure (s.t.p.)*.)

standard temperature and pressure (s.t.p.) is a pressure of 1.01325×10^5 Pa and a temperature of 273.15 K (or 0°C) exactly.

Chemists often choose 298.15 K (or 25°C) instead of 273.15 K (or 0°C) because when exact measurements are to be made of physical constants, it is much easier to control accurately a temperature which is slightly warmer than the surroundings than to cool a mass of gas, say, to 0°C before making the measurements.

standing wave: see *stationary wave*.

star: a large incandescent mass whose core pressures, densities and temperatures are high enough to sustain nuclear fusion reactions which provide a continuous and enormous output of radiant power. When the fuel begins to run out, other reactions cause the core to overheat and the star embarks upon the *red giant* phase.

The observable properties of stars include surface temperature T, distance D, mass m, luminosity L, intensity I, radius r (indirectly), velocity, spectrum and spectral type.

- Temperature is measured from the *total radiation curve* for the star using *Wien's law* (1893).

- Distance **D** can be measured directly only for those stars which are close enough to the Earth for their parallaxes to exceed 0.02 seconds of arc. Parallax is the maximum angle subtended at the star by the radius of the Earth's orbit. It is found by measuring the position of the star against a background of faint and generally distant stars over a six month period. The distances of a few thousand nearby stars have been measured in this way (i.e. about one star in a hundred million for our galaxy). The distance of a star for which the parallax would be exactly one second of arc is called a parsec (3.09×10^{16} m or 3.26 light years). The nearer stars are a few parsecs away.

- Mass can be determined for any star which is one member of a double star and close enough for the separation of the two stars to be measured. This is the case for about 200 stars.

- Luminosity L of a star is the total radiant power output in watts. If luminosity and surface temperature are both known, then using *Stefan's law*, the radius of the star can be found: $L = A\sigma T^4 = 4\pi r^2 . \sigma T^4$.

- The intensity I of a star is the radiant power per square metre of star surface, reaching the Earth's surface. If D is the distance of the star from the Earth, then $I = L/4\pi D^2$.

(Avoid the word 'brightness'. Luminosity is actual brightness, intensity is apparent brightness.)

Intensity is sometimes expressed in terms of two ancient parameters, the apparent magnitude m and the absolute magnitude M. Apparent magnitude is related to intensity I as follows:

$$m = -2.5 \lg I + \text{constant} \qquad \text{or} \qquad m_1 - m_2 = 2.5 \lg (I_2 / I_1)$$

The absolute magnitude of a star equals the apparent magnitude of the star were it a distance of ten parsecs from the Earth. Since intensity is inversely proportional to distance squared, we can substitute M for m_2 on the left hand side of the last equation and, on the right hand side, replace (l_2/l_1) by $(r/10)^2$. The equation now links m to M for a particular star:

$$m - M = 5 \lg (r/10)$$

where r is the distance of the star from the Earth in parsecs. These same two quantities, m and M, are used with galaxies as well as stars.

Spectral types (O, B, A, F, G, K, M) are a way a classifying the spectra of stars by broad visual character. The method was essential before Wien's law was known; it is retained because it is useful.

Class	Temperature	Surface character
O	> 25 000 K	Ionised helium lines (He II) dominant
B	25 000 K to 11 000 K	Neutral helium (He I), excited hydrogen visible
A	11 000 K to 7500 K	Hydrogen lines dominant
F	7500 K to 6000 K	Hydrogen lines weaker, calcium gaining strength
G	6000 K to 5000 K	Calcium dominant, ions much weaker
K	5000 K to 3500 K	Neutral metals predominate
M	< 3500 K	Molecular bands apparent

See *end states of stars, Hertzsprung–Russell diagram, Hubble's law, models of the universe, parallax (stellar)*.

states of matter are the set of states that the atoms or molecules in a lump of matter can exist in. At one time there were thought to be three states of matter: solid, liquid and gas. More recently the gaseous state has been joined by the plasma state, while the distinction between the solid and liquid states has broken down by the recognition that *glasses* form an intermediate group. It is not simply that there are kinds of solid, there is an increasing number of structures which do not fit the definition of the solid state as an ordered arrangement with the atoms or molecules attached to their neighbours. (See *solid, liquid, gas, metals, crystalline solids, polycrystalline solids, amorphous solids, polymeric solids*.)

stationary wave: an extended oscillation in which vibrational energy is stored. Stationary waves are sometimes called standing waves. Stationary waves can be resolved into two *progressive waves* of the same frequency, travelling along the same line but in opposite directions. The sequence of events leading to the formation of a stationary wave is illustrated in the diagram opposite.

An initial disturbance sends out waves in both directions along a taut wire. They are reflected from both ends of the wire and overlap on their return paths. At certain points *(nodes)* the waves meet always out of phase and interfere destructively. At other points *(antinodes)* the waves meet always in phase and interfere constructively. The wire is stationary at the nodes and oscillates with maximum amplitude at the antinodes.

The waves are stationary in two senses: there is no flow of energy and the node–antinode system is fixed in position. The final diagram illustrates what is meant by all points between adjacent nodes moving in phase and by a phase change of π on moving through a node.

Most musical instruments work by setting up a stationary wave on a thread, in a pipe or on a diaphragm. The energy arrives in an erratic way but is stored smoothly and is transmitted to the air in a smooth fashion. The dying away of the note from a single piano string is an example of *damped harmonic motion*.

A stationary wave is an *interference* pattern set up by two waves of the same frequency moving in opposite directions. It is also an example of *resonance*. Imagine a power source of variable frequency applied to a taut wire. Let the frequency slowly increase from zero. At most frequencies, the two waves moving in opposite directions along the wire will self-destruct. At some frequencies (the resonant frequencies) they enhance one another. This happens when a stationary wave can form on the wire with a node at either end. These waves are the result of constructive interference and are an example of resonance. Resonance can happen only on a wire the length of which is a whole number of half-wave-lengths. The situation with pipes is more complicated. (See *waves in pipes, waves on strings*.)

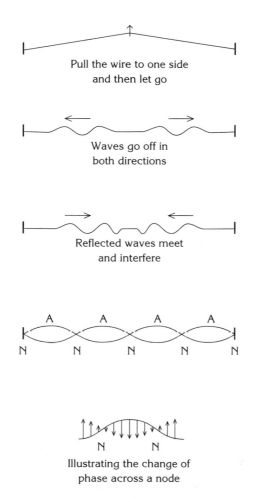

Pull the wire to one side
and then let go

Waves go off in
both directions

Reflected waves meet
and interfere

A   A   A   A
N   N   N   N   N

N   N
Illustrating the change of
phase across a node

The table on page 240 lists the important properties of stationary waves and contrasts them with progressive waves.

Stationary wave	Progressive wave
Stores vibrational energy	Transmits vibrational energy
The amplitude increases smoothly from a minimum value at the nodes to a maximum value at the antinodes	The amplitude is constant along the length of the wave
All points along the wave between any two adjacent nodes move in phase.	Phase varies linearly with distance along the path of the wave
There is an abrupt phase change of p on moving through a node	There are no sudden changes of phase
Nodes are set half a wavelength apart, antinodes are set half a wavelength apart, an antinode is set midway between two nodes.	There are no nodes or antinodes.

steady state is the condition of a body when the various temperatures throughout the body are constant, and during which there is no net loss or gain of heat to or from the surroundings (note however that the temperatures are not necessarily everywhere the same).

Imagine a long handled metal spoon dipping into water kept at boiling point. Heat energy flows into the bowl of the spoon and out along the handle. The spoon loses heat energy to the surroundings along its length and the top end of the handle will be cool enough to hold. The spoon loses as much heat as it gains, all the temperatures are steady but the temperatures are not everywhere the same. Furthermore, heat energy is flowing through the spoon without there being any net gain or loss. This is a steady state, not a *thermal equilibrium state*.

steam point: the temperature at standard atmospheric pressure of steam in *thermal equilibrium* with water. It is 373.15 K.

Stefan's law states that the power radiated (i.e., the energy radiated per second), W, from a body is proportional to the area, A, and to the fourth power of the absolute temperature of the surface, T. For a perfect radiator (a total radiator or a black body) Stefan's law is written:

$$W = \sigma A T^4$$

This equation gives only the heat radiated. The heat received from the surroundings must obey the same law, otherwise a body at the same temperature as its surroundings would not be in thermal equilibrium.

Take as an example a black pot in surroundings at 300 K. Raise the temperature of the pot to 600 K. The net power loss is the difference between the power radiated and the power absorbed:

$$W = \sigma A (600^4 - 300^4) = \sigma A\, 300^4\, (2^4 - 1)$$

Since 2^4 is 16, we see that a body which is twice the temperature of its surroundings emits 16 times as much radiation per second as it receives.

Stefan's law is extended to cover real surfaces by writing it in the form:

$$W = \varepsilon \sigma A T^4$$

where ε is a constant between 0 and 1. ε is called the total emissivity and it has a different values for different surfaces.

stiffness is the quality of a material which enables it to respond to a high tensile stress with a small tensile strain: its *Young's modulus* is large. If various materials are made up into beams, all the same size, the stiffer materials will bend least. A stiff material has low flexibility. (See *stress-strain curves*.)

Stokes' law states that the viscous force opposing the motion of a sphere of radius r moving at terminal speed v through a fluid medium of viscosity η in conditions of streamline flow is $6\pi r\eta v$. Imagine a small steel sphere, mass m and radius r, falling through a light oil in a tall measuring cylinder. Its speed increases slowly up to a maximum and constant value called the *terminal speed*. At this point the apparent weight of the steel sphere {= weight − upthrust}, pulling downwards, and the resistive drag, acting upwards, are equal and opposite. The resistive drag at the terminal speed v equals $6\pi r\eta v$. (See *streamline flow*, *viscosity*.)

straight line motion is the fundamental motion on which Newtonian mechanics and classical physics are built. Some books and specifications call it rectilinear motion. Kinematics, the study of motion without reference to force or energy, takes the topic beyond straight line motion to curved paths of almost any shape. Straight line motion in A level physics is covered by the *equations of motion*. One danger with straight line motion is that, because direction Is often irrelevant, the distinctions between distance and displacement and between speed and velocity become obscured. Once force or energy are involved, the topic becomes part of dynamics. (See *equations of motion*.)

straight wire: for a long straight wire isolated from its surroundings and carrying current I, the magnitude of the magnetic field strength B at a point a perpendicular distance r from the wire is given by:

$$B = \frac{\mu_0 I}{2\pi r}$$

The magnetic lines of force are all circular and the spacing increases away from the wire as the field strength weakens. The direction of the magnetic lines of force is shown in the diagram below. This direction is given by the *right hand grip rule*.

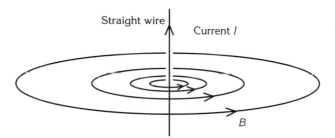

Remember when copying this diagram that the magnetic lines of force must be broken to show where they pass behind the wire and that you should draw a minimum of four magnetic lines of force to show the increased spacing with distance. (See *magnetic effect of current*.)

strain, or more accurately tensile strain, is the increase in length of a specimen per unit length in response to a tensile *stress*. Strain is a ratio of two lengths; it is a number with no unit. It is a property of a material and its value depends only on the magnitude of the applied stress and not at all on the dimensions of the specimen. See *stress*, *stress-strain curves*, *Young's modulus*.

streamline flow (or laminar flow) is the smooth flow of a fluid along a channel without the disturbing action of small eddies. The streamlines are the imaginary lines which mark the path of the fluid. In streamline flow these lines are smooth curves or straight lines. *Poisseuille's equation* and *Stokes' law* are valid for streamline flow only. (See *turbulence*, *Reynold's number*.)

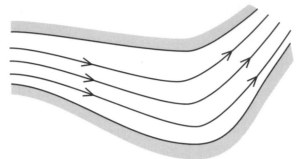

strength of a specimen is the tensile force or the tensile *stress* needed to break it. (See *ultimate tensile stress*, *breaking stress*.)

stress may compress a volume, distort a shape or stretch a thread. Only the stretching action matters at A level. The specimen must be pulled by equal tensions in opposite directions. Tensile stress, as it is called, equals the tensile force per unit area applied to a specimen. The SI unit for tensile stress is the pascal (Pa). See *strain*, *stress–strain curves*, *compressibility*, *Young modulus*.

stress–strain curves are graphs which show how the tensile *strain* of a specimen varies with the tensile *stress* applied to it. The graph in the first diagram applies to no material in particular; this is why there are no numerical values on the axes. Its purpose is to point out typical features of these curves and to define some of the parameters which are used when discussing the elastic properties of materials.

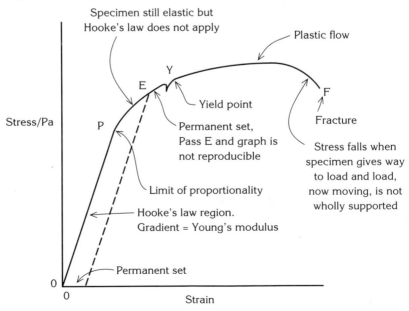

Starting at O, the point of zero stress and zero strain, the curve rises along a straight line to the point P. Along OP, stress is proportional to strain, and *Hooke's law* applies to the specimen. OP is called the Hooke's law region and P is called the limit of proportionality. The gradient of the straight portion OP is the *Young's modulus* for the specimen. Along the section PE, the specimen is still elastic in the sense that if the strain is removed the specimen will return to O and recover its original length. The point E is the elastic limit. If the specimen is stretched beyond E, it will not return to O when the stress is relaxed; it will never quite recover its original length. There is now a permanent set. Increase the stress a little more and the specimen is subject to sudden internal slippage. This is called the yield point. The curve drops and rises because with the sudden increase in length (increase in strain) the wire moves and does not altogether support the load for a moment. This part of the curve cannot be repeated or traced backwards. Once the yield point is passed, any further increase in the level of stress invites a continuum of small yields which is known as plastic flow. Atoms are slipping over one another, the specimen is about to break and since the load is no longer being supported (it is falling and gathering speed) the stress decreases. The area between the curve and the strain axis is the energy transferred per unit volume to the specimen during the whole process.

The set of curves below and on page 244 is chosen to highlight different elastic properties. Each curve starts at zero stress and continues until the specimen breaks.

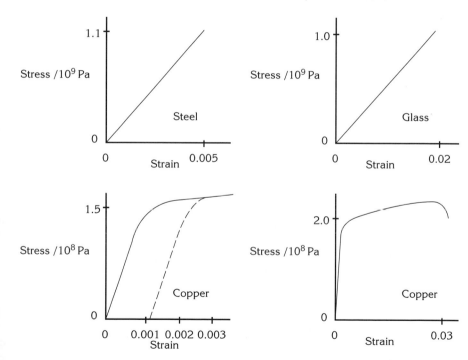

Mild steel and glass have high breaking stresses and are said to be strong. *Stiffness* is linked to the gradient of the graph and is higher for steel than for glass. Glass gives no warning that it is about to break, it is said to be brittle. Copper, with its large extension and large area 'under the curve', absorbs a lot of energy before it breaks and is said to be tough.

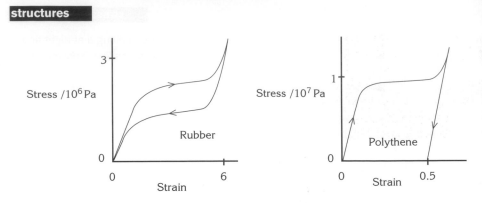

Copper exhibits a large plastic flow while rubber has a strong *hysteresis* effect. The numerical values on the axes of these curves are orders of magnitude; they should be known. (See *brittleness*, *toughness*, *plastic deformation*.)

structures is a term used to describe manufactured objects such as *beams*, girders, pipes, walls or embankments where the design process has made use of the physical properties of the component materials.

subatomic particle is a term which covers any particle smaller than an atom in size. Atoms are roughly 10^{-10} m across. Atomic nuclei are typically 10^{-14} m across unless the mass number is low, and then, like the neutron and proton, they are closer to 10^{-15} m across. *Fundamental particles* such as *leptons* (e.g. electrons) or *quarks* are known to be less than 10^{-18} m across. That is about as small as can be measured. A distance of 10^{-15} m is just within the range of operation of the strong nuclear force.

superconductors are substances with a resistance to current which falls effectively to zero a few degrees above absolute zero. Magnets powered by coils of superconducting materials are in widespread use already. Transmission of electric power by superconducting cable will await the development of materials which are modestly priced, available in bulk and retain their superconducting properties at temperatures higher than the boiling point of liquid air.

supernova: the first stage in the end state of a main sequence *star* whose mass is greater than eight solar masses. It is characterised by excessive luminosity and a comparatively short time scale. See *end states of stars*.

superposition is the process by which two waves combine into a single wave form when they overlap. (See *principle of superposition*.)

symbol: an agreed abbreviation for a physical or mathematical quantity. It may represent either a constant or a variable. See page 281 for a list of symbols used in physics for variables and constants.

systematic error: an inherent error in an instrument or in technique which cannot be reduced or eliminated by averaging the results of several measurements. Systematic error is guarded against by analysis and care. Examples include zero error, measuring the wrong distance for a pendulum length or a falling body, poor instrument calibration and failing to correct for background radiation. (See *error*, *uncertainty*, *random error*, *zero error*.)

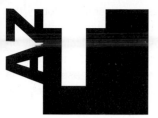

telescopes magnify by collecting light confined to a small visual angle and spreading it out over a much larger angle. In terrestrial terms, the same effect is achieved by walking closer to the object of interest. If you get too close to the object for the eyes to focus, then you throw the image back and away from the eye with a magnifying glass or, still better, a microscope. The ray diagram for an astronomical telescope in normal adjustment (i.e. with the image at infinity) is shown in the diagram.

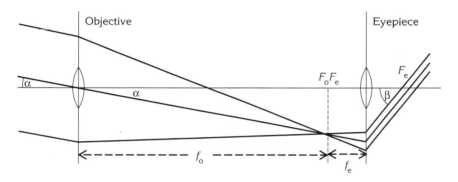

The magnification is calculated as follows:

$$M = \frac{\text{angle subtended by image at eye}}{\text{angle subtended by object at eye}} = \frac{\beta}{\alpha} = \frac{f_o}{f_e}$$

The image is never perfect . Two principal sources of imperfection are *chromatic aberration* (change of focal length with wavelength of light used) and *spherical aberration* (change of focal length with distance from the principal axis). The elimination of chromatic aberration is a pleasing feature of reflecting telescopes but the main attraction of reflectors has to do with resolving power.

The angle θ between two point sources of light which can just be separated by a telescope can be shown to equal $1.22\lambda/D$. where D is the diameter of aperture which limits the width of the light beam collected and λ is the wavelength of the radiation used. In practice, D is the diameter of the objective lens or of the mirror. The angle θ is called the resolving power; it governs how much detail can be seen. Mirrors can be made much larger than lenses without them distorting under their own weight. For high resolution, reflectors win every time. The long wavelength of radio waves explains the huge size of radio telescopes. See *diffraction at a small circular hole, resolving power*.

The second diagram shows the optical system for a reflecting telescope. Without the small central mirror or something similar, the light cannot leave the telescope and the image will not be seen. The light loss associated with the mirror is hugely compensated for by the larger diameters possible with reflectors. The focusing of either refracting or reflecting telescopes can be adjusted to give real images which can then be observed electronically on PC monitors using *charge-couple devices* (CCDs) as image sensors. With radio telescopes, the small mirror is replaced by an aerial tuned to the radio frequency the astronomer is interested in, usually around 1.5 GHz. For resolving powers well below one degree, the radio reflector needs a diameter of 100 m or more. Resolution can be improved by using arrays of smaller radio telescopes. The downside to this is a loss of collecting beam cross-section.

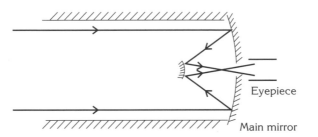

The reason for the concentration of work within the visible and radio wavebands is that these are the wavelengths for which the atmosphere it transparent. Mounting telescopes on satellites extends the work to other wavebands such as X-ray, UV, IR and microwave. There is the further advantage in using satellites, which applies equally to the visible region, that there is little atmospheric absorption or refraction.

temperature is the hotness of a body expressed on a numerical scale. Temperature scales were, at one time, set up in an arbitrary way and different people used different scales. It is now recognised that temperature is not an arbitrary quantity. The *absolute thermodynamic temperature scale* is defined within thermodynamics in a way which is independent of the properties of any substance but applicable to all substances. It turns out that this scale is identical to the temperature scale that is written into the gas laws. Using these ideas as a basis, another temperature scale has been defined – the *International Practical Temperature Scale* – which defines how temperatures are to be measured before they can be part of a legally enforceable specification.

One advantage of basing everything on thermodynamics is that there is no need to make assumptions about mercury in a mercury thermometer, platinum in a resistance thermometer or any other substance. Each of these, if assumed to vary linearly with temperature, would define its own temperature scale. This is an altogether different thing from the *Celsius* and Fahrenheit scales; these just involve shifting the zero of the absolute scale or changing the temperature interval belonging to one degree or both.

temperature coefficient of resistance is the increase in the resistance of a conductor, as a fraction of its resistance at 0°C, per degree absolute rise in temperature. The word equation definition is:

$$\text{temperature coefficient of resistance} = \frac{\text{increase in resistance}}{\text{resistance at 0°C} \times \text{temperature rise}}$$

$$\alpha = \frac{R_2 - R_1}{R_0(T_2 - T_1)} \ K^{-1}$$

temperature gradient is defined by the word equation:

$$\text{temperature gradient} = \frac{\text{temperature difference between two points}}{\text{distance between the points}}$$

The unit for temperature gradient is $K\ m^{-1}$.

tempering involves heating a specimen to below red heat thereby making it tough and springy. It is applied after quenching to remove the brittleness caused by the *quenching*. (See *annealing*.)

tensile force is a stretching force usually applied to a material that is in the form of a strand such as a wire, an elastic string or a spring. For the material to stretch, there must be two tensile forces of equal magnitude, acting in opposite directions. The ratio of force to extension depends both on the material and its dimensions. To remove the dimensional factor, it is usual to work with (tensile) *stress* and (tensile) *strain*.

terminal speed is the speed of fall of a body in air for which the air resistance, acting upwards, is equal in magnitude but opposite in direction to the downwards acting weight. The resultant force is zero and the body continues its path towards the ground at constant speed. For a large but light body, such as a parachutist, the Archimedean upthrust has to be taken into account. It acts upwards and 'helps' the air resistance. There are then three forces drawn and labelled on the free-body force diagram.

Terminal speeds in liquids are much slower than terminal speeds in air. The Archimedean upthrust is generally a much larger fraction of the weight and the drag forces supplied by the liquid are much larger, than those provided by air.

An important special case is illustrated by a solid sphere falling through liquid at rest. Provided that the liquid is not held in a narrow tube, the drag force is given by *Stokes' law*. It equals $6\pi a \eta v$ where a is the radius of the sphere, η is the coefficient of viscosity and v is the terminal speed. Stokes' law provides a method of measuring coefficient of viscosity and, if the value of the coefficient is known already, a method of estimating average particle size.

tesla is the SI unit for magnetic field strength. The unit symbol is T. Following on from the equation used to define magnetic field strength, $F = B\,I\,l$, the tesla is defined by the word equation:

$$\text{tesla} = \frac{\text{newton}}{\text{ampere metre}}$$

(See *magnetic field strength*.)

thermal conduction is the process by which heat flows through matter along a temperature gradient. The mathematics for three-dimensional flow quickly gets out of hand so we restrict ourselves to the one-dimensional flow of heat along a uniform bar of cross-sectional area A. We further restrict ourselves to the steady state condition, wherein temperatures along the bar, though different, are all constant. The situation is illustrated in the diagram on page 248.

One end of the bar is kept at 100°C with a current of steam, the other at 0°C with a current of ice-cold water. The bar is in a steady state once all the temperatures along its length are constant, (this is not a *thermal equilibrium* state, in which all the temperatures along its length would be the same and there would be no heat flow). If the sides of the bar are efficiently lagged, the rate of heat flow into the hot end of the bar equals the rate of heat flow

247

out of the cold end. The rate of heat flow will be the same at all points along the length of the bar and the temperature gradient $d\theta/dx$ is constant along the length of the bar. The graph shows the temperatures at different distances from the hot end of the bar as a straight line. The heat flow is governed by the equation:

$$\frac{dQ}{dt} = kA\frac{d\theta}{dx}$$

where dQ/dt is the heat flowing per second into or out of the bar, A is the area of cross-section of the bar, $d\theta/dx$ is the temperature gradient and k is the coefficient of thermal conductivity of the material of the bar. The unit for k is W m^{-1} K^{-1}. This equation defines the coefficient of thermal conductivity, k.

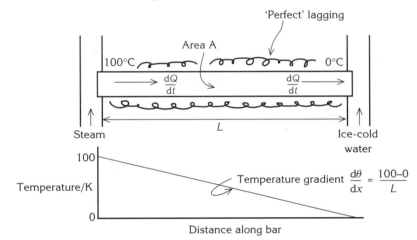

Worked example

Imagine that the bar shown in the diagram is made of brass (thermal conductivity 109 W m^{-1} K^{-1}), that its length is 25.6 cm and its cross-sectional area 9.2 cm^2. The temperature difference between the ends is 100 K. Find the rate of flow of heat along the bar.

$$\text{rate of flow of heat} = \frac{dQ}{dt} = kA\frac{\theta_1 - \theta_2}{x}$$

$$= (109 \text{ W m}^{-1} \text{ K}^{-1})(9.2 \times 10^{-4} \text{ m}^2)(100 \text{ K} / 0.256 \text{ m})$$

$$= 39 \text{ W} \quad \text{(to 2 s.f.)}$$

(See *thermal conductivity, temperature gradient, U value*.)

thermal conductivity is the rate of heat flow per unit area per unit temperature gradient. Its value is a property of the substance involved. The symbol used is k and the SI unit is W m^{-1} K^{-1}. The word equation definition is derived from the thermal conductivity equation thus:

$$k = \frac{dQ}{dt} \left/ A\frac{d\theta}{dx} \right.$$

$$= \frac{\text{rate of flow of heat}}{\text{area} \times \text{temperature gradient}}$$

$$\text{W m}^{-1} \text{K}^{-1} = \frac{\text{J/s}}{\text{m}^2 \times (\text{K/m})}$$

There are two groups of substances; good thermal conductors and poor thermal conductors. The values of the coefficient k for good thermal conductors are typically a thousand times as great as the values of k for poor thermal conductors. Most good thermal conductors are electrical conductors, (i.e. metals). The reason is that the conduction electrons carry heat through the material far faster than do atomic vibrations within the solid mass. Compare this with the electrical case: crystal structures do not transmit electrical charge. The distinction is between conductors and insulators and the ratio of their electrical conductivities is typically around 10^{21}.

Building materials need to be good thermal insulators, but they need other characteristics such as strength, rigidity, elasticity, transparency and to be proof against noise, wind, weather, frost and rain. The best compromises are mostly found in *composite materials*.

thermal energy is an uncertain term which is best avoided. That part of the energy content of a body which is associated with the disordered motion of the atoms (kinetic in part and potential in part) is called *internal energy*. *Heat* is a flow of energy down a temperature gradient. There are four mechanisms for *heat transfer*: radiation, conduction, convection and evaporation. Heat conduction has always been discussed under the heading *'thermal conduction'*.

thermal equilibrium is a state in which every part of a body has the same temperature and in which there is no net loss or gain of thermal energy by the body to or from the surroundings. A body is known to be in a state of thermal equilibrium if the independent physical properties of the system are constant, (pressure and volume). Once other observable physical properties are known to be constant, the temperature of the body must be constant.

Two bodies will be in thermal equilibrium with one another (their temperatures equal) provided (i) they are in thermal contact and (ii) they are both in thermal equilibrium states. Do not confuse a thermal equilibrium state with a *steady state*.

(See *zeroth law of thermodynamics*.)

thermal fission reactor (nuclear reactor): a power plant which exploits the energy released when nuclei of heavy elements such as uranium are broken down into lighter elements. The mass-energy conversion process which underpins the process is discussed under *'binding energy'*. The principle of a thermal fission reactor involves three on-going activities: a stable chain reaction, an energy removal process and maintenance.

Stable chain reaction: start with a stray neutron which collides with a uranium nucleus. The nucleus absorbs the neutron, it becomes unstable and breaks up into two nuclei, X and Y, some neutrons (two or three but averaging 2.6) and a lot of energy (mostly in the form of kinetic energy for X and Y). We can write a generalised nuclear equation for one fission process:

$$n + {}^{235}U = X + Y + 2.6n + \text{about 200 MeV}$$

This is not as dangerous as it looks. The incident neutron must be 'slow'. That is, its kinetic energy must correspond to the prevailing temperature. But the 2.6 neutrons are 'fast'; their kinetic energies are five or six orders of magnitude too high for them to trigger further fission reactions.

To achieve a stable chain reaction, the 2.6 fast neutrons must be slowed down, with one neutron returning to cause a successful fission reaction and the rest absorbed or escaping from the structure. This slowing down process occurs in the *moderator*. The uranium fuel is held in long narrow fuel rods which slot into columns of graphite moderator at regular intervals. When fast neutrons escape from the fuel rods into the moderator, they crash into carbon nuclei and give up most of their energy after several such collisions. The moderator gets hot, a proportion of the neutrons diffuses out of the system but some neutrons drift back at random into the fuel rods. The number drifting back is just large enough to sustain a constant rate of nuclear fission processes (see the diagram below).

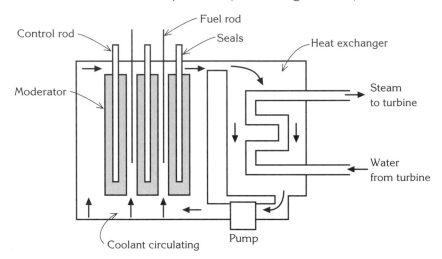

The stability of the chain reaction is a consequence of the design, the amount of uranium in the fuel rods and the spacing of the fuel rods in the moderator. There is one more vital component: a set of control rods. These control rods contain material such as boron which is highly efficient at absorbing neutrons. The rods are lowered into the moderator at regular intervals and just deep enough for the reactor to 'go critical'. If they are lowered further the reaction slows down; if they are raised slightly the reaction speeds up.

The energy removal process uses a fluid, possibly carbon dioxide gas under high pressure, pumped through the reactor. The coolant picks up by conduction the heat generated in the fuel rods and transmits it to the heat exchanger where it powers the conversion of water into high pressure steam used to drive turbines and generate electrical power. Since most of the energy released in the fission processes appears as kinetic energy of the heavy fission products, the coolant pipes must be as close as possible to the fuel rods, otherwise the moderator material will overheat. Notice that the fission products, excepting the neutrons, stay in the fuels rods. The neutrons cannot pick up the kinetic energy from these fission products: they are too light and they are already far too fast so that collisions would tend to slow them down. Only a conduction process can remove the energy from the fuel rods and for this the coolant must not be far away.

The maintenance work includes replacing the fuel rods at regular intervals and monitoring the coolant, coolant support system, moderator and control rods.

thermal power is power drawn from any device for which the starting point is the burning of some kind of fuel. See *thermal power station*.

thermal power station: an installation where fuel is burned and part of the heat released, used to generate electrical power. The overall efficiency of a thermal power station is the ratio of the electrical energy reaching the consumer to the internal energy released when the primary fuel burns.

thermionic emission is the flow of electrons through the surface of a hot metal and away. The metal has to be heated to a dull red heat and a battery is needed to sustain the flow; otherwise the build up of positive potential on the metal inhibits and eventually suppresses the electron flow. A heated cathode provides the electrons in the electron guns used in X-ray tubes, oscilloscopes and television sets.

thermistors are resistors with strongly temperature-dependent resistance values. The current I in a conductor of cross-sectional area A is given by the relationship:

$$I = nAqv$$

where there are n charge carriers per unit volume moving with an average drift velocity v and each carrying charge q. For some semiconducting materials, the number of charge carriers per unit volume, n, increases sharply with rising temperature. For a given potential difference across the specimen, the current increases sharply with increasing temperature. The resistance will have dropped. These thermistors have a negative *temperature coefficient of resistance*; they are the most common sort.

Other thermistors can be manufactured in which the increase of n with rising temperature is slight but in which lattice vibrations impose a much stronger brake on charge carrier motion if the temperature rises. There is a sharp decrease in v and the resultant fall in current swamps any tendency for the current to rise because of an increase in n. The resistance of these thermistors increases if their temperature rises: they have positive temperature coefficients of resistance.

thermocouple: a thermometer which uses the *thermoelectric effect* to sense and measure a difference in temperature. The diagram below shows an idealised thermocouple. A series circuit is set up consisting of a digital voltmeter, two wires of metal X and a single wire of metal Y. The junctions at either end of metal Y, labelled A and B, are set in two objects of temperatures θ_A and θ_B respectively. Experiment shows that the reading of the digital voltmeter depends on $(\theta_A - \theta_B)$. If θ_B is fixed at 0°C by immersing the end B in iced water, the digital voltmeter can be calibrated to give a direct reading of the temperature θ_A.

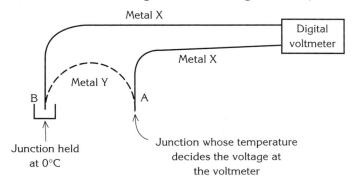

Junction held at 0°C

Junction whose temperature decides the voltage at the voltmeter

Thermocouples are not easy to build because their calibration depends on too many factors, such as the purity of the metals used for the wires and the quality of the junctions. The metals used often need special solder. In commercial thermocouples, the metals are chosen to suit the temperature range for which they will be used.

Thermocouples are widely used because of their many valuable qualities:

- they are rugged, accurate and sensitive
- they are usually small enough not to interfere with the object whose temperature is being measured
- output can be fed directly to a data recording device
- the instrumentation can be remote from any hot object under investigation.

thermodynamics (first law of): the first law states that any increase ΔU in the internal energy of a system is the sum of the heat ΔQ flowing into the system and the work ΔW done on the system:

$$\Delta U = \Delta Q + \Delta W$$

Internal energy correlates with temperature. Heat is energy flow driven by a temperature gradient and work is ordered energy transfer.

A number of points can be made.

- Energy is defined as the physical quantity whose conservation is governed by this law. Consequently, the first law of thermodynamics is a formal statement of the law of *conservation of energy*. It is an assertion and cannot be tested by experiment. Part of the reason for this is the way in which ΔU is defined – but we can't follow this matter any further at this level.

- The law hints that there is some fundamental difference between energy ΔQ and energy ΔW. This is indeed the case. ΔQ is energy associated with molecular motion and is wholly random and disordered. ΔW is energy associated with something directed, like a force or a voltage difference; it is an ordered property. The second law of thermodynamics teaches us that only a fraction of the disordered energy could ever be recovered from the system in a manner which does useful work. But the ordered energy, if it is stored in an ordered fashion, can all be converted to useful work.

- There can never be a physical situation in which the three quantities, ΔU, ΔQ and ΔW, can be measured separately. The first law of thermodynamics cannot be checked by experiment.

When tackling a question on the first law, remember that a positive or negative value for ΔU is indicated by a rise or fall of system temperature or by a change of state (melting, freezing, etc.). If neither effect is present then ΔU is zero, the system is possibly in a thermal equilibrium state, certainly in a steady state, and ΔQ must equal $-\Delta W$. Three possibilities are nicely exemplified by an electric lamp immediately after it has been switched on. The electric power source does work on the filament $\Delta W = IV\Delta t$. At the instant of switching on, the filament is at the same temperature as the surroundings and ΔQ must be zero. It follows that ΔU equals ΔW and the temperature of the filament begins to rise. While the filament temperature is rising, ΔQ is growing in a negative sense (energy radiated from the filament) and there is a corresponding fall in ΔU, the energy which stays behind to raise the filament temperature. Eventually a steady state is reached; ΔU is zero, and ΔQ is $- \Delta W$. The radiant energy leaving per second equals the work done per second on the filament by the electrical power source.

thermoplastic polymers are *polymeric solids* with molecular chains that are weakly bound together by *Van der Waals' forces*. Such polymers soften on heating and harden again on cooling. They are easily moulded into shape.

thermosetting polymers are *polymeric solids* in which strong *covalent* cross-links are formed between the molecular chains when the material first cools but which do not break up when the material is reheated. They are good for making saucepan handles and suchlike, but can be brittle.

threshold frequency: the minimum frequency of electromagnetic radiation capable of triggering the *photoelectric effect* for a given material.

thrust is a force spread over an area instead of acting through a point of application. The thrust exerted by a fluid on a surface equals the product of the *pressure* and the area of the surface. It is a contact force which acts normal to the surface and into the surface.

time is a base physical quantity in SI. Its symbol is t and the SI unit is the *second*. The time scale that clocks measure and which we use without too much fuss in mechanics and in the rest of A level physics is defined implicitly within Newton's laws of motion. See *base quantity*.

time constant τ is the time interval during which the charging or discharging current for a capacitor in series with a resistor falls to a fraction $1/e$ of an earlier value. The phrase 'an earlier value' is used because the current–time graph is exponential and the time needed for any given fractional change is constant along its length.

The charging current for capacitance C in series with resistance R changes with time according to the following formula:

$$I = I_{max}\, e^{-t/RC}$$

The definition of time constant tells us that when $t = \tau$,

$$I = I_{max}\, e^{-1}$$

Clearly, for $t = \tau$,

$$\tau/RC = 1 \quad \text{or} \quad \tau = RC$$

This last equation is an important result. But it is not the definition of time constant. The definition of time constant is the first sentence of this entry; learn it.

time division multiplexing is a technique for enabling one telephone cable or transmission channel to carry many separate messages in digital form simultaneously. Imagine the time scale for such cable to be split into intervals, each one microsecond long. Next, imagine each of these one microsecond divisions to be divided into 20 slots, each of length 50 ns. One digital message can be transmitted in the third slot of each microsecond, another can be transmitted in the 14th slot. If one slot is retained for managing the use of the cable, 19 messages can be transmitted simultaneously. (See also *frequency division multiplexing, pulse code modulation*.)

time interval is a measure of duration. It is measured in seconds.

torque is the moment of a couple. The SI unit for torque is the newton metre (N m). The symbol used is T. It is a *pseudo-vector quantity* but its vector qualities are usually ignored at

A level (these vector properties matter only if a body is acted upon by two non-parallel torques). Torque plays the same part in rotational motion that force plays in translational motion. The vital relationships are:

torque = moment of inertia × angular acceleration

$$T = I \times \alpha$$

torque × angle turned in radians = increase in rotational kinetic energy

$$T \times \theta = \tfrac{1}{2} I \omega_2^2 - \tfrac{1}{2} I \omega_1^2$$

torque × angular speed = power

$$T \times \omega = P$$

The word torque is often used in an informal sense to mean a 'turning force' as opposed to a push or a pull.

Worked example

A torque of 0.20 N m is applied to a stationary flywheel of moment of inertia 1.42 kg m². Calculate the speed of the flywheel at the end of the fifth revolution.

θ after 5 revolutions is 10π radians. ω_1 is zero, whence:

$$\omega_2 = \sqrt{\left(\frac{2T\theta}{I}\right)} = \sqrt{\left(\frac{2 \times (0.20 \text{ N m}) \times (10\pi \text{ rad})}{(1.42 \text{ kg m}^2)}\right)} = 3.0 \text{ rad s}^{-1} \quad \text{(to 2 s.f.)}$$

(See *moment of inertia, angular acceleration*.)

total internal reflection occurs when light approaches a refracting surface at such an angle that there is no angle of refraction that would satisfy Snell's law. This is best illustrated with a worked example.

Worked example

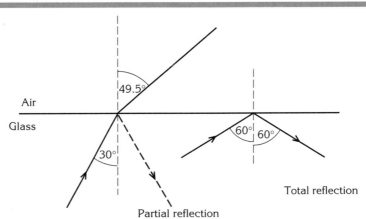

The diagram shows two rays of light in glass travelling towards a flat surface, one with an angle of incidence of 30° and the other with an angle of incidence of 60°. Find the angle of refraction in air corresponding to the angle of incidence in glass of 30°.

$$\frac{1}{1.52} = \frac{\sin 30°}{\sin r} = \frac{0.5}{\sin r}$$

$$r = \sin^{-1}(1.52 \times 0.5) = \sin^{-1}(0.76) = 49° \quad \text{(to 2 s.f.)}$$

But sin 60° is 0.8660 and r would come out at $\sin^{-1}(1.52 \times 0.8660)$ or $\sin^{-1}(1.32)$ and there is no angle whose sine has this value. Light in glass approaching the flat surface at an angle of incidence of 60° cannot be refracted.

Since it cannot be refracted all the light is reflected. This is called total internal reflection. The light which approaches the surface at 30° is partially reflected and partially refracted. Between these two angles there is a so-called critical angle which marks the onset of total internal reflection. The critical angle is shown in the diagram below.

The critical angle c must be the angle for which r is $\sin^{-1}(1)$. But r is $\sin^{-1}(1.52 \times \sin c)$. We have then:

$$1.52 \times \sin c = 1 \quad \text{or} \quad c = \sin^{-1}\left(\frac{1}{1.52}\right) = 41° \quad \text{(to 2 s.f.)}$$

More generally, the critical angle c within a material of refractive index n relative to vacuum, is calculated from the relationship:

$$c = \sin^{-1}\left(\frac{1}{n}\right)$$

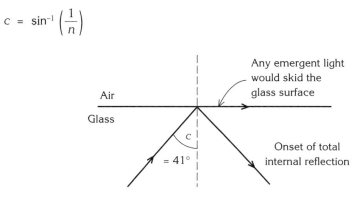

total radiation curve (more often 'black body radiation curve') is a graph showing how W_λ, the power radiated per unit area per unit wavelength interval from a hot surface, varies with wavelength (see the graph on page 256).

The phrase 'per unit wavelength interval' can be a little troublesome. It means this. Measure the power carried away from the surface within the wavelength band λ to $\lambda + d\lambda$ and divide this amount of power by $d\lambda$. Then make $d\lambda$ as small as possible. It is just another example of a *limiting value*.

There are a few points to make about the graph on page 256.

The word 'total' emphasises that all wavelengths are represented. The term 'black body' emphasises that perfect radiators (surfaces which give the smooth curve shown with nothing missing) are black when cold. That is, they are not only complete radiators, they are also perfect absorbers.

Two curves are shown, one corresponding to a higher temperature than the other. The surface at the higher temperature emits radiation more powerfully than the surface at the lower temperature: this is true for all wavelengths but it is exaggerated in the short wavelength region.

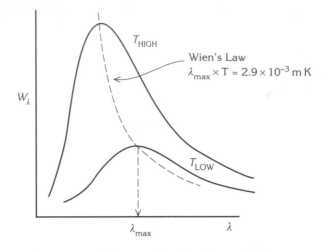

The area between the curve and the wavelength axis equals the power radiated from unit area of the surface over the whole wavelength range. This area increases in proportion to T^4. This is *Stefan's law*.

The wavelength, λ_{max}, at which the curve peaks is proportional to $1/T$. That is, it gets shorter as the temperature of the radiating surface increases. This is called *Wien's law*.

The graph is a way of illustrating the *electromagnetic spectrum*. To complete this task, the high temperature would have to be identified, say 6000 K, and the wavelength axis divided into sections and labelled UV, visible, IR and so on.

toughness is a measure of the amount of *work* which is needed to break a material. Materials such as leather are tough. It contrasts with *stiffness* and *brittleness*.

tracer elements: *radioactive* isotopes used to monitor the progress of an element or its compounds through a living organism or process pathway (such as in the soil or through a gas pipe) by providing non-radioactive isotopes of the same element with a radioactive 'tag'.

transformer: two coils wound on a single core thereby enabling electrical power to be transmitted between two circuits which are electrically isolated from one another.

The diagram shows an idealised transformer. The primary coil has N_p turns, the secondary coil N_S turns. The primary coil is used to supply energy to the transformer, the secondary coil to carry the energy away. The theory is simplified by assuming that coils are perfect conductors and that there is no leakage of *magnetic flux* from the transformer core.

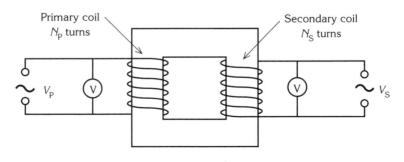

Assume, as a starting point, that an alternating current in the primary coil creates a changing magnetic flux $d\phi/dt$ within the core. There will be an induced e.m.f. E in every turn of wire in either coil where

$$E = -\frac{d\phi}{dt}$$

The primary coil consists of N_p turns of wire in series and the induced e.m.f. will be N_pE. This induced e.m.f. is equal to and in the opposite direction to the alternating voltage V_p applied to the primary coil by an external power source.

The secondary coil consists of N_s turns of wire in series. The induced e.m.f. (the only e.m.f.) will be N_sE. This will be the voltage across the secondary coil, V_s. We have, then:

$$V_p = -N_pE = N_p\frac{d\phi}{dt}$$

$$V_s = N_sE = -N_s\frac{d\phi}{dt}$$

$$\frac{V_p}{V_s} = -\frac{N_p}{N_s}$$

The negative sign indicates that the primary and secondary voltages are in antiphase.

There is the additional requirement that any power drawn from the secondary coil must be provided by the power source for the primary coil. If I_p and I_s represent the primary and secondary currents respectively, then:

$$I_pV_p = I_sV_s$$

(See *mutual inductance*.)

translation motion: the motion of a body which moves from one point to another instead of undergoing a repetitive motion such as simple harmonic motion or circular motion.

transverse waves are *progressive waves* whose displacements are at right angles to the direction of energy flow. (See *longitudinal waves*, *stationary waves*.)

triangle of forces: this states that if three coplanar forces are in equilibrium, then the vectors representing the forces, if laid end to end, will form a triangle. This is one condition for three coplanar forces to be in equilibrium. The second condition requires the lines of action of the three forces to meet at a point. The two conditions together imply that each of the three forces is equal and opposite to the resultant of the other two.

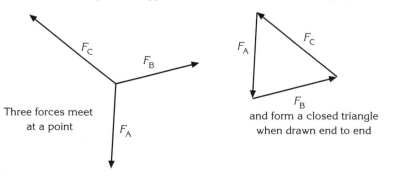

Three forces meet at a point

and form a closed triangle when drawn end to end

triple point: the one temperature at which ice, water and water vapour, all in their pure form, may be in *thermal equilibrium* with one another. (See *ice point*.)

turbulent flow is the churning motion in a fluid which occurs when *streamline flow* has broken down. It happens when a fluid is travelling too fast or when it passes an obstacle in its path. (See *Reynold's number*.)

turning is the motion of a body such that all parts of the body move along concentric circular paths in a plane at right angles to the axis of rotation. If the axis of rotation lies outside the body, the motion is called circular motion and needs a *centripetal force* to sustain it. If the axis passes through the body, then it is called *rotational motion*. Rotational motion, once established by an unbalanced *torque*, is self-sustaining because *angular momentum* is a conserved quantity.

two-dimensional motion describes the many situations in which a body moves along a curved path in a plane. *Projectiles* provide a common example. Problems are usually tackled by treating the two velocity components separately. An exception is *circular motion*: this is a two-dimensional motion but problems are usually tackled from a different standpoint. The two components of circular motion would, as it happens, be two simple harmonic motions 90° out of phase, along paths at right angles to one another and bisecting one another.

two slits experiment: a demonstration of the wave nature of light by showing the *interference* pattern where two coherent beams of light meet on a screen. The diagram illustrates the minimal set up. The two slits are parallel, a distance s apart, in an opaque sheet. These slits are illuminated normally by light from a laser. The light which gets through the slits illuminates a screen a perpendicular distance D beyond the slits.

The pattern on the screen consists of a dozen or so equally-spaced bright lines. They are parallel to the slits in the opaque sheet and their centres are a distance x apart. The central bright line, A, is equidistant from both slits in the opaque screen. The beams of light from the two slits reach A in phase. There is constructive interference and the screen is bright. Each point B on either side of A is one wavelength further from one slit than the other. The beams of light from the two slits arrive at the screen in phase. There is constructive interference and the screen is bright at the points B.

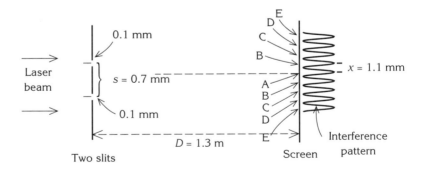

Midway between A and either point B, one beam travels half a wavelength further than the other, the beams are out of phase when they meet and the interference is destructive.

The points C are two wavelengths further from one slit than the other, the points D are three wavelengths apart and so on.

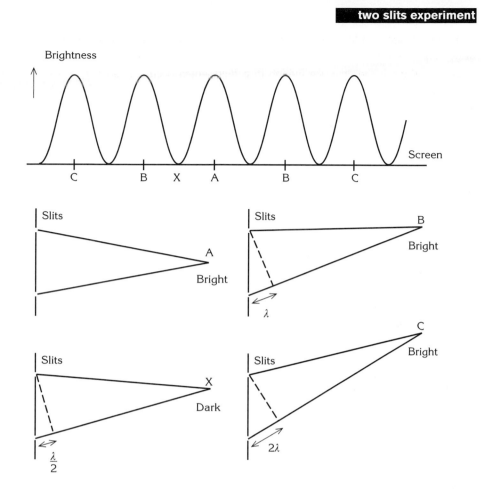

It can be shown that, if λ is the wavelength of the laser light, then:

$$\lambda = \frac{xs}{D}$$

In a typical experiment, the slits are about a tenth of a millimetre wide and about 0.7 mm apart. Distance D would be about 1.3 m and x about 1.1 mm. The wavelength λ of the laser light would be around 600 nm. Values of s and x would be measured with a micrometer eyepiece, D would be measured with a ruler.

The slit width (≈ 0.1 mm) is of some significance. If it is too large, the interference pattern loses definition, if too small, the pattern is sharp but too faint to see. This is why the best images are recorded photographically; a long exposure time brightens the sharp but too-faint image

Consider the part played by diffraction in the formation of the interference pattern. There are two diffraction patterns on the screen, one for each slit. These patterns are each a couple of centimetres wide but only s apart at their centres (in this case, 0.7 mm). The envelope formed by merging these two diffraction patterns decides the brightnesses of the different bright fringes in the interference pattern. Briefly, the two diffraction patterns overlap and interfere with one another, constructively along the bright fringes and destructively in between. Consequently, the central fringe is the brightest and the brightnesses of the other fringes fall away on each side. This falling away of brightness has not been shown on either diagram. See *diffraction at a single slit*.

Another way of looking at this experiment is to regard the slits as point sources of light in the plane of the diagram. (In fact, Thomas Young did use pin holes as point sources, not slits.) There will be constructive interference at any point on the screen which is one or more whole wavelengths further from one slit than the other. A little calculation shows that these points are a distance $\lambda D/s$ apart and this gives us the equation for the distance x between the centres of adjacent fringes. Destructive interference occurs at any point which is an odd number of half wavelengths further from one slit than the other. These points of darkness are mid-way between the points of maximum brightness. This treatment explains how the bright and dark bands are formed and leads to the expression for fringe width, but it does not explain why the bands extend only a centimetre or so on either side of the principal axis. This point is discussed in detail under *diffraction at a single slit*.

If you insist on writing out the explanation in terms of crests and troughs, then get it right! Constructive interference happens where crest meets crest and destructive interference where crest meets trough.

Historically, the important feature of this experiment is the dark lines, the destructive interference. They mark out parts of the screen which would be illuminated by either slit on its own but are in darkness when illuminated by both slits at the same time. The observation that one beam of light can cancel another was altogether original.

Do you need revision help and advice?

Go to pages 290–306 for a range of revision appendices that include plenty of exam advice and tips.

ultimate tensile stress is the tensile *stress* at which 'necking' begins. Necking is a disproportionate localised deformation which develops into a region of excessive *strain* prior to breaking.

ultrasound is high frequency ($\approx$ few MHz), short wavelength ($\approx$ 1 mm) sound used for non-destructive exploration of the inside of the human body. The two basic ideas are:

1   the distance of a reflecting surface from a transmitter/receiver is the journey time of a pulse multiplied by the speed of the ultrasound

2   the reflection coefficient of a surface is defined and calculated as follows:

$$\alpha = \frac{\text{reflected intensity}}{\text{incident intensity}} = \left(\frac{Z_2 - Z_1}{Z_2 + Z_1}\right)^2$$

where Z, the specific acoustic impedance of a medium, is the product of the density ρ of the medium and the speed c of ultrasound within the medium, $Z = \rho c$. The value of Z is typically 1.6×10^6 kg m^{-2} s^{-1} for soft tissues but changes as much as 20% in passing from say fat to muscle and increases by a factor of four at a tissue/bone boundary.

The strong dependence of reflection coefficient on density change explains the necessity for the coupling gel between the probe and the skin surface.

The ultrasound transmitter, usually built around a piezoelectric crystal, also acts as the receiver. The circuitry has four functions:

●   to generate the signal which becomes the pulse

●   to detect and amplify the reflected pulse

●   to record the time lapse between transmission and reception and

●   to display the result on a screen.

Display may be an oscilloscope trace showing receiver signal intensity (y) against time (x). This is called an A-scan. Alternatively the oscilloscope trace may be a straight line whose length indicates time and whose brightness indicates signal strength. This is called a B-scan. The advantage of the B-scan is that it can be built up into a two-dimensional picture of, say, a fetus within the mother's womb.

Another area of use depends on measuring the frequency difference between the transmitted pulse and the reflected pulse after reflection by a moving surface. The surface may be part of a patient's heart, the heart of a foetus or blood corpuscles carried along in the blood stream. The relevant speed equals the speed of the ultrasound in the medium on the near side of the reflector, multiplied by the fractional change in frequency ($\Delta f/f$) and divided by twice cos θ where θ is the angle between the direction of the reflector motion and the direction of the ultrasound.

ultraviolet light: see *UV*

uncertainty is the range of values on either side of a measurement within which the true value is expected to be located. There is however no more than a probability that the true value lies within this range. This is why some people prefer the term 'probable error'.

Imagine that you want to know the diameter of an old halfpenny and that all you have is a plastic ruler and a magnifying glass. The diameter looks just less than 25.5 mm but you are not too sure about how much. You can write 'diameter = (25.4 ± 0.1) mm' or you might prefer the form, 'diameter = 25.4 mm $\pm 0.4\%$'. The $\pm$ sign is characteristic of uncertainties; *errors* are either positive or negative. An uncertainty is a range of values, an error is one particular amount.

Imagine a hundred careful people making a hundred such measurements with your ruler. The hundred values will lead to a more precise statement. Some measurements may be as high as 25.6 mm and some will be as low as 25.2 but the average value about which all these measurements centre can be decided with much greater certainty (i.e. a smaller uncertainty) as something like (25.39 ± 0.02) mm.

Individual measurements above and below the mean value occur randomly and this contribution to the uncertainty is described as a *random error*. A feature of a random error is that it can be reduced by averaging several independent measurements.

Imagine, next, that a friend repeats the experiment using an old ruler that has shrunk over the years. The hundred measurements may yield a result like (25.43 ± 0.02) mm. The random error is unchanged but all the measurements are 0.04 mm too large. This is called a *systematic error*. Its cause is a fault in the instrument and no amount of repetition will indicate its size or reduce it. (See *precision*.)

uncertainty principle: a statement that the product of the uncertainty ΔE in the energy of a body and the time interval Δt for which it has this energy is less than or equal to $h/2\pi$, where h is the Planck constant. This principle touches A level physics in the particle physics area. An *exchange particle* can come into being for a time so short that the uncertainty principle is not violated. This works out at about 10^{-24} s for the weak interaction. Light can travel no further than about 10^{-16} m in this time interval. This explains why the weak interaction is a short range force which acts within, but not between, neutrons and protons.

unified atomic mass constant: the modern unit for what used to be called the atomic mass unit is the 'unified atomic mass constant'. It is defined as one twelfth of the mass of a carbon–12 atom and the symbol used is 'u' where 1 u = $1.660\,540 \times 10^{-27}$ kg. The energy equivalent ($\Delta m\,c^2$) is $1.492\,419$ J or 931.494 MeV. The unwieldy name of this unit distinguishes it from an older unit, the *atomic mass unit*, which was defined as one-sixteenth of the mass of an oxygen–16 atom.

uniform electric field: an electric field with magnitude and direction that are everywhere the same. The diagram on page 263 shows the uniform electric field between the plates of a parallel plate capacitor. Notice the 'edge effects' at each side. The field lines are equally spaced. So too are the lines of equal potential – the equipotentials – which are shown as broken lines.

If you draw this diagram in an examination, never have fewer than three equipotentials or five electric field strength lines. Be sure to have the spacing approximately correct.

The relationship between electric field strength, E, and rate of change of potential with distance, $- dV/dx$, is simpler for a uniform field. In general:

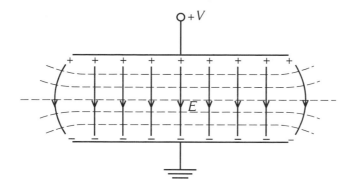

$$E = -\frac{dV}{dx}$$

But for a uniform field:

$$E = -\frac{dV}{dx} = -\frac{V}{d}$$

where d is the distance, measured parallel to, and in the same direction as E, along which the drop in potential is V. It is because V diminishes along the direction of E that the negative sign appears. Often, the direction in which d is measured is unimportant. Only the magnitudes are of interest. We then write:

$$E = \frac{V}{d}$$

unit: an agreed amount of physical quantity used to express the result of a measurement. It has to be assumed that instruments are available which can compare the magnitudes of two different amounts of the physical quantity. A ruler, for instance, will tell you that one length is 5.65 times as long as another length. If the shorter length is one metre then the greater length is 5.65 m.

The magnitude of any physical quantity is expressed as the product of a number and a unit.

The SI system defines seven *base* (physical) *quantities* and their corresponding seven *base units*. All other units are defined in terms of these seven base units and are called *derived units*.

The SI unit for acceleration is the m s^{-2}. The m s^{-2} is a derived unit; acceleration does not have units, it has a unit. The unit of force is kg m s^{-2} and this unit has a name of its own, the newton (N). Some derived units have their own names, most do not.

upthrust is a term which means any upward thrust but is usually retained to describe the upward force acting on an object immersed in a fluid. (See *Archimedes' principle*.)

U value is the rate of heat flow through a thermal barrier per unit area, per unit temperature difference across the barrier. The symbol is U and the SI unit W m^{-2} K^{-1}. The word equation definition for U value is:

$$\text{U value} = \frac{\text{rate of heat flow through a thermal barrier}}{\text{area} \times \text{temperature difference}}$$

Compare the U value equation with the *thermal conductivity* equation:

$$\frac{dQ}{dt} = kA\frac{d\theta}{dx} = UA\, d\theta$$

In purely mathematical terms, we should be able to define U as the quotient k/dx. But this would obscure the fact that k is a property of a homogeneous material while U is a property of a structure such as a window or a wall treated as a thermal barrier. U values are used in the building trade.

Worked example

A small detached house has 220 m² of double cavity walling (U value 0.51 W m⁻² K⁻¹) and 21 m² of double glazing (U value 4.8 W m⁻² K⁻¹). Calculate the ratio of the heat losses through the windows to the heat losses through the walls.

$$\text{ratio of heat losses} = \frac{[U \times A]_{\text{windows}}}{[U \times A]_{\text{walls}}} = \frac{(4.8 \text{ W m}^{-2} \text{ K}^{-1})(21 \text{ m}^2)}{(0.51 \text{ W m}^{-2} \text{ K}^{-1})(220 \text{ m}^2)} = 0.90$$

(See *thermal conduction*.)

UV or ultraviolet light is that segment of the *electromagnetic spectrum* which is emitted when the outer electrons of atoms in excited states return to their ground states, provided that their wavelengths are too short to be visible. Typical UV photons have energies in the region of several electron volts and this is more than sufficient to break up many atomic bonds. It follows that UV radiation is chemically active in a destructive sense and very damaging.

UV radiation consists of *photons* with energies just low enough for them to be present in the sun's spectrum. Most of the energy in the UV waveband carried in the solar spectrum is absorbed by ozone in the stratosphere, so protecting life on Earth from its damaging effects. This also explains why the temperature of the stratosphere increases with height. This increase of temperature with height eliminates convection currents and largely contains weather within the troposphere (the lowest 20 kilometres or so of the atmosphere).

The UV segment of the electromagnetic spectrum ranges from about 1 nm to about 400 nm where it meets the visible spectrum. The corresponding frequencies range from about 0.3 EHz (0.3×10^{18} Hz) down to about 0.75 PHz (0.75×10^{15} Hz) and the corresponding photon energies range from about 1.24 keV to about 3 eV. The high frequency end of the sun's spectrum is close to 10 PHz (10×10^{15} Hz; a photon energy of around 40 eV); photons of this energy, though few in number, would be very damaging if they reached the Earth's surface.

UV radiation is detected mostly photographically or with photocells. Many substances fluoresce when irradiated with UV. (See *photoelectric effect*.)

Van der Waals' bond is a weak intermolecular bond. An atom consists of a positive nucleus surrounded by electron shells. When isolated the atom is electrically neutral and creates no electric field. But in the liquid or solid state, an atom can be so distorted by the close presence of its neighbours that the 'centre of charge' of the electrons is displaced away from the nucleus. The atom becomes a dipole, that is, a positive charge and an equal negative charge separated by a short distance.

The negative pole of the distorted atom would be attracted to a positively charged component of a neighbouring molecule; the positive pole would be attracted to negatively charged components. Neighbouring atoms or molecules are similarly distorted and so tend to stick to the atoms in a neighbouring molecule. The net result is an inter-atomic bond which is much weaker than an *ionic* or *covalent* bond but which is strong enough to be significant. This is particularly so with polymeric solids in which a single polymer molecule can be held down by hundreds of Van der Waals' bonds along its length. Surface tension and viscosity begin with Van der Waals' bonds.

vector quantity: a physical or mathematical quantity which has both direction and magnitude. Examples include displacement, velocity, acceleration, force, magnetic field strength and electric field strength. Force is different from other vectors: it acts through a point. (For a discussion of force as a vector, see *force: a localised vector*.) The phrase 'composition of vectors' means no more than how vectors add together. Two vectors must represent the same physical quantity before they can be added together.

Vector quantities are often represented by arrows. The length of an arrow represents vector magnitude and the direction of the arrow represents vector direction. Forces are a little more difficult; the arrows have to point along the correct lines of action. (See also *parallelogram law*, *equilibrant*, *resultant*, *resolution of vectors*.)

velocity is defined by the word equation:

$$\text{velocity} = \frac{\text{change of displacement}}{\text{time interval}}$$

The unit for velocity is m s^{-1}.

The first graph on page 266 illustrates the ideas of uniform velocity and uniform acceleration, the second illustrates average velocity and instantaneous velocity.

The term 'uniform velocity' implies equal changes of displacement in equal times. Since velocity is a *vector quantity*, velocity can only be uniform for straight line motion. The term 'uniform acceleration ' means that the velocity increases by equal amounts in equal times without changing direction.

The magnitude of a velocity equals the *gradient* of a *displacement–time graph*. Look at the second graph, average velocity during the period t_A to t_B is the gradient of the straight line AB and instantaneous velocity at time t_c is the gradient of the tangent at C.

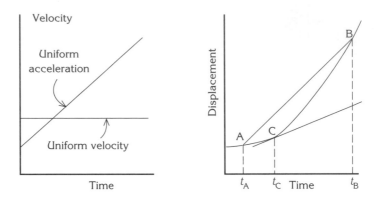

Average velocity during a specified time interval is defined by the word equation:

$$\text{average velocity} = \frac{\text{change of displacement}}{\text{time interval}}$$

Instantaneous velocity is defined as the limiting value of the ratio on the right-hand side of this equation evaluated for a very short time interval.

The *equations of motion* apply to motion along a straight line with uniform acceleration. Velocity changes in magnitude but not in direction.

For uniform circular motion, speed is constant but velocity changes continuously. The magnitude of the velocity is constant but its direction turns uniformly through 360° during each complete circulation. It accelerates, but maintains a constant speed. Similarly, the magnitude of the acceleration is constant but the direction is towards the centre of the circle at all times. It, too, turns uniformly through 360° during each complete circulation.

(See *displacement, acceleration, limiting value, straight line motion, velocity–time graph*.)

velocity–time graph: a graphical display of the time variation of *velocity*. Velocity is a *vector quantity* and its value can be positive or negative; the graph can cut the time axis. The graph cannot indicate the changing direction of a moving body – if the body is following a curved path of some sort, then each component of the velocity must have its own velocity–time graph.

The area between the velocity–time curve and the time axis equals change of displacement during the relevant time interval and this can be positive or negative. Each graph is linked to a particular direction and the gradient of a tangent to the graph gives *acceleration* in that direction for that instant. It may be positive or negative.

The first velocity–time graph on page 267 is for a car which is driven away from the observer (reference point) along a straight line with increasing acceleration. The gradient of the graph is the instantaneous acceleration and the area between the graph and the time axis is the change of *displacement* during the relevant time interval.

Notice that the area between the curve and the time axis is change of displacement and not just displacement. It will equal displacement if the area starts at an instant when the displacement is zero.

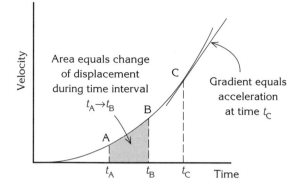

Worked example

A glider on a horizontal air track travels towards one end at a speed of 0.60 m s⁻¹. It is reflected elastically and returns along the same path at the same speed. The clock is started when the glider is 1.2 m from the reflector and is stopped when the glider returns through the same point. Draw and label a velocity–time graph for the motion.

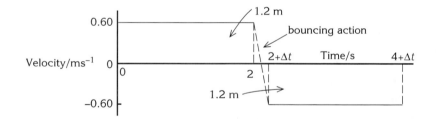

Notice that the average velocity calculated over the 4 + Δt seconds time interval is zero and the net displacement is zero.

viscoelasticity is the liquid-like flow within a polymer or glass under the action of a long-term continued stress. The effect is more noticeable at high temperatures than at low. The separate effects of the time-dependent element and the stress-dependent element are illustrated with bouncing putty. Under the action of the fast forces involved in bouncing, it is almost perfectly elastic. But under slow continued pressure it moulds quite easily in the hand and is almost perfectly plastic.

viscosity is the frictional force which opposes movement within a *fluid*. It is what stops tea whirling round a cup forever once it has been stirred. It acts between two adjacent layers of fluid which are moving parallel to one another but at different speeds. There are materials such as toffee or glass which have no clear-cut melting point: they get softer as the temperature rises. The viscosity has so high a value at the lower temperatures that the liquid material behaves in many ways like a solid. The Earth's mantle is a fluid with a high viscosity and this is why the drift of continents is so slow. Honey and treacle have much higher viscosities when cold than when hot. Gases, too, have viscosity.

The flow of a viscous fluid is analysed with reference to a coefficient of viscosity. To understand how the coefficient of viscosity, η, is defined, look at the situation in the diagram below.

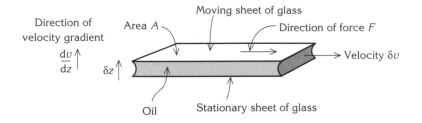

A thin flat sheet of oil of area A and thickness δz lies between two sheets of glass. The tangential force F needed to keep the upper sheet moving across the lower sheet with a steady speed δv is given by the word equation:

tangential force = coefficient of viscosity × area × normal velocity gradient

$$F = \eta A \frac{dv}{dz}$$

Notice that the force F is parallel to the area A and that the velocity gradient is measured at right angles to A. This equation, which defines coefficient of viscosity, is valid for *streamline flow* only. Any small eddies (known as *turbulence*) within the fluid between the two sheets of glass have a dramatic effect on the magnitude of the force F.

Using this definition, two important results can be derived. The first, *Poisseuille's equation*, applies to the flow of fluid along pipes. The second result, *Stokes' law*, applies to a sphere falling through a fluid at its terminal speed.

visible spectrum: the segment of the *electromagnetic spectrum* which can be detected by the normal human eye. The wavelength range is from close to 400 nm to close to 700 nm. The corresponding frequencies are 750 THz (7.5×10^{14} Hz) to 430 THz (4.3×10^{14} Hz). The *photon* energies range from 3.1 eV to 1.8 eV and are just high enough to ensure that visible light has some destructive power at the molecular level and this is why some dyes fade so quickly in bright sunlight. Sunlight should be treated with some caution – nothing extravagant, just a little care and good sunglasses in strong sunlight.

The principal source of visible radiation is the sun, a large radiating mass at a temperature of 5780 K. Most of this energy is radiated within the IR segment but about 20% of it is within the visible segment of the electromagnetic spectrum. As the temperature of a body rises, the onset of visibility is round about 1000 K (750°C). The peak of the total radiation spectrum moves into the visible region at about 4000 K.

Visible radiation is emitted when excited or ionised atoms return to their ground states in stages. For the direct transition the radiation would be in the UV range. The precise wavelength values associated with these transitions are the main indicators science has to atomic structure. This is the origin of the study known as spectrum analysis.

volt (V): the SI unit of *potential difference*. It is another name for the joule per coulomb.

voltage–current characteristics: see *current–voltage characteristics*.

voltmeters measure potential difference. Digital voltmeters are accurate and cheap. There is no point wondering how they work unless you have a special interest. They must be connected across the required two points and in parallel with the rest of the circuit. The diagram below shows how a voltmeter and ammeter should be connected when measuring the current in, and the potential difference across, a resistor R.

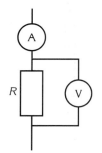

vulcanised rubber is rubber which is hardened by mixing the raw rubber with sulphur and then exposing it to a slow heating process. The sulphur binds the polymer molecules together by providing cross-links. The hardening which results is in rough proportion to the concentration of sulphur. (See *rubber*.)

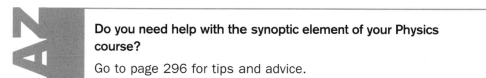

Do you need help with the synoptic element of your Physics course?

Go to page 296 for tips and advice.

water waves are waves which run across the surface of a mass of water. Water waves in nature are of two kinds: the large, imposing waves which move across the sea (gravity waves) and the tiny waves seen typically running across a puddle in a light breeze (ripples). Gravity waves are driven by the wind and provide a major route for energy loss from the atmosphere. They are called gravity waves because the restoring force which accounts for the oscillatory character of the motion is gravity. Ripples, whose speeds are typically about a fifth of a metre per second, are wind driven too but the restoring forces are controlled by the surface tension of water. Were these ripples to travel much faster than they do, *ripple tanks* would lose their usefulness.

watt: a rate of energy transfer of one joule per second. It is the SI derived unit for *power* and its symbol is W. The *base unit* equivalent of the watt is $kg\ m^2\ s^{-3}$.

wave behaviour is the set of properties (*reflection*, *refraction*, *diffraction*, *interference*, *dispersion*) shared by all wave motions (and *polarisation phenomena* which is shared by some). The diagrams below summarise how most of these phenomena can be demonstrated with a ripple tank.

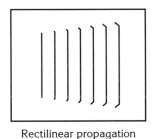

Rectilinear propagation

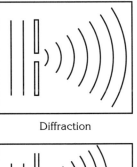

Diffraction

Refraction

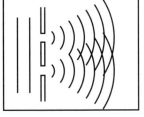

Interference

wavefront: a surface which travels with a wave, drawn everywhere at right angles to the direction of the wave and across which the phase of the wave is everywhere the same.

wavelength: the shortest distance along the path of a wave between two points where the wave displacements are changing in phase. (See *sound waves*, *electromagnetic waves*.)

wave nature of light: the capacity of a light beam to behave like a succession of waves progressing through a medium. Light shares its wave-like properties with the whole of the *electromagnetic spectrum* but the phrase 'wave nature of light' pre-dates the discovery of the rest of the electromagnetic spectrum apart from the near infra red and the near ultra violet. The key properties which betray the wave nature of light are *interference* and *diffraction* and the relationship of refractive index to wave speed. *Polarisation* establishes the transverse nature of these waves. The straight line motion of light and its reflection by polished surfaces can be reconciled with the wave idea but sits more easily with a particle interpretation (see *wave–particle duality*).

Wave theory cannot accommodate those properties of light which link up with its interaction with matter and with the *photon* concept. These properties are principally the *photoelectric effect*, the formation of absorption and emission spectra and the character of the *total radiation* (black body) spectrum.

wave–particle duality expresses the need to treat both photons and atomic or sub-atomic particles as either particles or wave trains, according to circumstances. Electrons act mostly like particles but they can be diffracted. Electromagnetic radiation is known best for its wave properties but the photoelectric effect underlines the particle nature of this radiation.

wave speed: See *speed of a wave*.

waves in pipes are *stationary waves*. They form at particular frequencies only. This is because their formation is an example of *resonance*. There is a resonant frequency when the air in a pipe is set into vibration with a node at any closed end of the pipe and an antinode at any open end of the pipe. If n is the number of antinodes within the pipe, l is the length of the pipe and λ is the wavelength of the wave-motion, then, for a pipe open at one end and closed at the other (as in the diagram),

$$l = (2n + 1)\frac{\lambda}{4}$$

Notice that $(2n + 1)$ equals 1 or 3 or 5 etc. ... it is always an odd number. For resonance, the length of the pipe must be an odd number of quarter wavelengths.

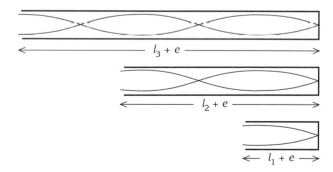

In practice, the position of the exterior antinode is a distance e beyond the open end of the pipe. The equation relating the length of pipe to the wavelength of the sound needs modifying slightly:

$$l + e = (2n +1)\frac{\lambda}{4}$$

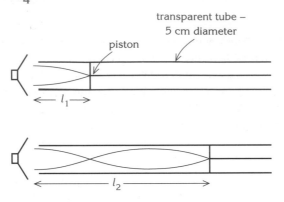

In the arrangement shown above, the position of the piston is adjusted for resonance. The frequency, f, applied to the loudspeaker is held constant and the two shortest resonance lengths, l_1 and l_2 are measured. Since:

$$l_2 - l_1 = \frac{\lambda}{2}$$

once these two pipe lengths are measured, the wavelength of the sound from the loud-speaker can be calculated without the end correction e having to be known. If the frequency of the sound is known, the speed of sound in the pipe can be found. (Remember that this experiment is not accepted as a method of measuring the *speed of sound* in free air.).

waves on strings are *stationary waves*. They form when a string is set into vibration with a node at each end of the string. If n is the number of antinodes, l is the length of the string and λ is the wavelength of the wave motion on the wire, then, since nodes are half a wave-length apart:

$$2l = n\lambda$$

But wavelength equals speed/frequency ($\lambda = c/f$) and the speed of waves along a string is $\sqrt{(T/\mu)}$ where T is the tension in the string and μ is the mass per unit length of the string. It follows that a stationary wave will form on a string at those frequencies for which:

$$f = \frac{n}{2l}\sqrt{\left(\frac{T}{\mu}\right)}$$

Taut strings will support stationary waves at these frequencies and at no others. This is what is meant by *resonance*.

weight is the force acting on a mass due to the gravitational field in which it is placed. The SI unit of weight is the newton (N).

weightlessness is the condition of a body when the only force acting on it is its *weight*. This is the case for a satellite high up above the Earth's atmosphere. If you jump into the air, you will be near enough weightless until you touch the ground again.

white dwarf: the end state of main sequence stars whose masses are less than eight solar masses. See *end states of stars*.

Wien's law states that the wavelength λ_{max} at which the power radiated from a total radiator peaks is related to the surface temperature T of the radiator as follows:

$$\lambda_{max} T = 2.9 \times 10^{-3} \text{ m K}$$

The law enables the temperature of a bright but distant object to be determined spectroscopically. (See *total radiation*.)

wire is metal in the form of a thread. There are one or two features that affect the choice of wire used in a circuit. The current that a wire will carry without undue heating depends on its cross-sectional area. Power cable in the home is thicker and more expensive than lighting cable. Wire leading to an electric cooker needs to carry about 50 A and it is thicker still. The cooker must not be connected to the ring main, nor must an electric shower. The cables carrying electricity into the home are as much as 2 cm thick.

If the wire needs to be flexible, it is constructed of a large number of thin wires with a total cross-sectional area equal to the required value. A good example is the thick, many stranded wire used to connect an amplifier to the loudspeakers in a stereo system. A high quality stereo system uses a greater total thickness of wire and with more strands per unit area; it costs more but the voltage drop across the wire is significantly smaller so there is negligible signal loss in the cable.

Wire should be attached to three-pin plugs with some care. The end of the multi-stranded cable needs twisting on itself before it is curled round the threaded shaft ready for the nut to be tightened down on it. There are four possible combinations of twist direction and curl direction. Only one of them is right. Remember to make sure that the outer casing of the cable is firmly grasped at the entry to the plug.

The 'life' of electric cable is generally decided by the quality of the insulation. The PVC cladding of domestic wiring is reckoned to last about 25 years. Many appliances fail because the mechanical flexing of a power lead at the point where it enters the appliance causes the insulation to wear and fail at that point.

word equation: an equation in which the mathematical symbols are replaced by the names of the variables they correspond to. Word equations are frequently used for definitions of physical quantities.

work is the amount of energy transferred to an object when it is moved by a force applied to it by a different object. It is a *scalar quantity*. The SI unit of work is the joule (J). The symbol for work is W, the same as for energy. Amount of work and the joule are defined by word equations:

    work = force × component of displacement in direction of force

    joule = newton × metre

This definition shows that the joule is a convenient name for the newton metre.

Newton's third law of motion states that forces occur in equal and opposite pairs acting on different bodies. The force a bow exerts on an arrow equals the force the arrow exerts on the bow. But there is an essential lack of symmetry. The arrow moves in the direction of the force applied to it and it receives (kinetic) energy. The bow releases (potential) energy and does not have to move. Because energy is transferred from the bow to the arrow, it is the bow which works on the arrow and not the other way round. *Conservation of linear momentum* demands that the bowman receives a slight backwards push.

So far we have dealt with work and mechanical energy; let us enlarge the discussion to include other forms of energy. We started with the idea that work is an energy transfer process, and so it is. Heating is an energy transfer process but heating is not working, so what exactly is the difference?

The fundamental energy transfer principle is the *first law of thermodynamics* which is usually stated in the abbreviated form:

$$\Delta U = \Delta Q + \Delta W$$

ΔU is energy transferred to a system and it can arrive in two forms, ΔQ and ΔW.

Energy included in ΔQ is disordered energy like the kinetic energy stored in a gas.

Energy included in ΔW is ordered energy like the mechanical energy stored in a spring or the electrical energy stored in a capacitor. Transferring disordered energy is called 'heating'. Transferring ordered energy is called 'working'. The distinction is vital in thermodynamics because disordered energy can only be partly recovered in a useful way but ordered energy can be fully recovered.

Sometimes, as in frictional heating, the energy changes form on transfer. The object which provides the energy exerts a force, and so does work. Frictional heating contributes to ΔW and not to ΔQ. Similarly, electrical heating IVt contributes to ΔW and not ΔQ.

Next, how is an amount of work calculated? The three cases which follow relate to a force applied to a system (like a spring being stretched), a gas expanding against some resistance from the system (like compressed air in an air gun driving out a slug) and a battery driving electric current through a resistor (as in an electric fire).

- Mechanical work: force × component of the displacement parallel to the force.
- Gas expansion work: pressure × increase in volume.
- Electrical work: voltage × current × time.

Worked examples

1. A 3.2 kg mass falls through 1.6 m. How much work is done on it?

   Work done W = force × displacement = $(3.2 \text{ kg}) \times (9.8 \text{ m s}^{-2}) \times (1.6 \text{ m})$

   = 50 J (to 2 s.f.)

   The work is done by the gravitational field (that is where the energy comes from). The immediate destination for this energy is probably the kinetic energy of the falling mass.

2. A force of 56 N pushes a heavy box 1.9 m across a floor. Calculate the work done.

   $W = (56 \text{ N}) \times (1.9 \text{ m}) = 106 \text{ J}$

   Notice that in the first example, 50 J is probably stored as recoverable kinetic energy and $\Delta U = \Delta W$ for the falling mass. But in the second example the energy transferred by the working process is probably transformed into heat by friction. It is still part of ΔW in a first law of thermodynamics application because the energy provider is using ordered energy. It is stored in the receiving system as disordered (internal) energy and so contributes to a rise in temperature.

3. Estimate the work done when 60 cm³ of air at atmospheric pressure is compressed to 58 cm³.

$$W = -p\Delta V = (1 \times 10^5 \text{ Pa}) \times (2 \times 10^{-4} \text{ m}^3) = 20 \text{ J}$$

For this process, $\Delta U = \Delta W$. The energy arrives in the ordered form. Whether or not it is transformed to disordered energy in the gas, in whole or in part, is another matter. Notice that the work done on the gas is positive and the volume change is negative.

4. A 9 V battery drives a current of 0.4 A through a resistor for 3 minutes. Calculate the work done by the battery.

$$W = IVt = (0.4 \text{ A}) \times (9 \text{ V}) \times (180 \text{ s}) = 648 \text{ J}$$

Notice, again, that the energy provided by the battery is ordered energy (it can drive a motor or heat a resistor) so it will be part of ΔW and not ΔQ.

5. A 5 V battery charges a 56 µF capacitor. Calculate the work done by the battery and the energy stored in the capacitor. Explain the discrepancy.

charge stored in capacitor $Q = CV = 280 \text{ µC}$

work done by battery $= QV = (280 \times 10^{-6} \text{ C}) \times (5 \text{ V}) = 1.4 \text{ mJ}$

energy stored in capacitor $= \frac{1}{2}QV = 0.7 \text{ mJ}$

Half the energy from the battery is needed to drive current round the circuit. It is transformed into thermal energy in the resistive component of the circuit.

work done in compression: ΔW for a small change in the volume of a gas equals $-p\Delta V$, the product of the average pressure p and the change of volume $-\Delta V$. One consequence of this result is that the work done compressing a gas through a proportionately large change of volume equals the area between the line tracing the change on a p–V graph and the volume axis. A cylinder and piston arrangement is sometimes drawn parallel to the volume axis to illustrate what happens during the change. Notice that the pressure axis must go down to $p = 0$ if the area on the graph is to equal work done.

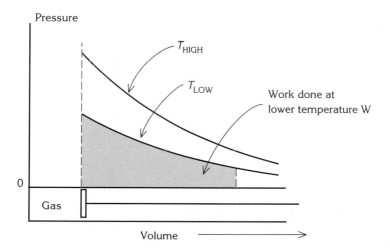

(See *work*.)

work function: ϕ joules, is the energy needed by a conduction electron to escape through the surface of a conductor. (See *photoelectric effect*.)

work hardening is the hardening or strengthening of a wire by subjecting it to repeated loading or *stress*, causing cold plastic deformation. It is a consequence of the increased number of tiny dislocations introduced into the metal.

What other subjects are you studying?

A–Zs cover 18 different subjects. See the inside back cover for a list of all the titles in the series and how to order.

X-rays need to be studied in terms of production, properties, detection and medical applications.

Production The essential components of an X-ray tube are an *electron* gun and a tungsten anode at opposite ends of an evacuated tube with a high voltage (e.g. 50 kV) source connected across them. A narrow beams of electrons streams across the tube onto the tungsten anode and X-rays are produced by the collisions. Most of the energy in the electron beam heats up the tungsten target; a small motor rotates the target to keep one spot from overheating. The ratio of X-ray energy output to electrical energy input (i.e. the tube efficiency) is very low because most of the electron beam energy is absorbed in the tungsten target, increasing its temperature.

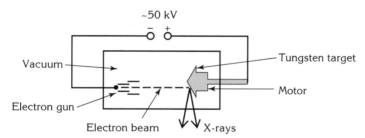

The second diagram shows a graph to illustrate how the X-ray intensity varies with X-ray photon energy for a tube voltage of 50 kV. Increasing the tube voltage widens the graph, increasing the tube current heightens it. The maximum photon frequency, f_{max}, is calculated from the relationship:

$$eV = hf_{max} = \text{kinetic energy of electron at anode}$$

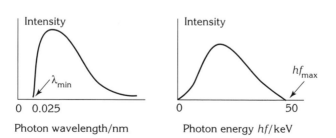

Should the electron beam energy be sufficient to excite electrons in the inner K-shells of the target material, then a line spectrum in the form of spikes is added to the continuous spectrum.

X-rays

Properties X-ray photon energies range from about 0.12 keV at the low end up to about 1.2 MeV at the high end. Wavelengths range from about 10 nm down to about 1 pm.

X-rays share all the properties of electromagnetic radiation. Because X-ray wavelengths are much the same size as atomic diameter, diffraction effects can be observed only by using crystals or thin sheets of metal as the diffracting elements. This leads into a major technique for examining crystal structure. Interference and polarisation effects are not of much relevance with X-rays but they have been observed. The short wavelengths of X-rays ensure that the inverse square law is followed closely.

Detection The two principal means of X-ray detection are photography and the GM tube. GM tubes are used when little more than an estimate of X-ray intensity is needed, usually for safety monitoring purposes. Photography gives a permanent record. Accurate work is mostly carried out with scintillation counter – photomultiplier assemblies which can both count the number of photons and record their energies (see *radiation detectors*).

Medical applications The medical uses are of two kinds: diagnostic and therapeutic.

a   Diagnosis. X-ray photographs of bones or lung tissues are used to identify fractures or other troubles. X-ray tubes for this work are run at a few tens of thousands of volts and the X-ray photons of this energy are largely absorbed in photoelectric events. Since the probability of these events increases with Z^3, where Z is the atomic number of the absorbing material, absorption in bones is strong but absorption in the soft tissues weak.

The quality of the X-ray photograph lies in its clarity and this improves with the intensity of the X-ray beam. But high beam intensity is bad for the subject of the photograph. To achieve a satisfactory image with a low intensity beam, the unexposed film is mounted between two fluorescent sheets which act as image intensifiers. Each X-ray photon which is absorbed in the screen triggers the emission of as many as two or three hundred visible photons (of energy a few eV). The advantage is a useful X-ray picture with a reduced X-ray dose. The cost is the blurring of the picture. In practice, a compromise has to be found.

b   Therapy. X-ray photons produced at around a megavolt or above are mostly absorbed by Compton scattering and this effect is independent of Z. There is no preferential absorption by any one tissue type. A beam of these X-rays, if focused upon some offending tissue, will destroy innocent tissue in its path. But if at the same time the body is rotated about the offending tissue, this will be the only tissue which gets a high enough dose to be destroyed. See *electromagnetic spectrum*, *photon*.

yield point: the point along a *stress–strain curve* at which random and erratic movement occurs between planes of atoms within the specimen. It marks the onset of *plastic flow*.

Young modulus is defined by the word equation:

$$\text{Young modulus} = \frac{\text{tensile stress}}{\text{tensile strain}}$$

$$\text{unit for Young modulus} = \frac{\text{pascal}}{\text{no unit}} = \text{pascal}$$

For a specimen in the form of a wire of length l and area of cross-section A subjected to a tensile force F which causes an extension e, the Young modulus E is calculated from the expression:

$$E = \frac{Fl}{Ae}$$

A rough order of magnitude can be found by putting typical values into the expression. With $F = 50$ N, $l = 3$ m, $A = 0.50 \times 10^{-6}$ m^2, and $e = 0.003$ m, then $E = 1 \times 10^{11}$ Pa.

Young's two slits experiment: see *two slits experiment*.

zero error arises when the scale reading of a voltmeter or ammeter is significantly different from zero when zero current flows through the instrument. It can also happen that the needle zeroes correctly but needs more than a finite current or voltage to shift it away from the zero. A zero error can occur with any instrument which has a scale: a thermometer, a micrometer screw gauge, a pressure gauge, an electrometer. Zero error cannot be reduced by averaging the results of several measurements: it is a *systematic error*. (See *error, uncertainty, random error*.)

zeroth law of thermodynamics: this states that if two bodies are each in *thermal equilibrium* with the same third body then they are in thermal equilibrium with one another. There must be some physical quantity which decides whether or not two bodies can be in thermal equilibrium with one another. This physical quantity is called *temperature*. The zeroth law of thermodynamics is a formal way of defining temperature with reference to the concept of a thermal equilibrium state. The attraction of the thermal equilibrium state concept is that it can be recognised without recourse to thermometers.

PHYSICAL QUANTITIES AND THEIR UNITS: BASIC AND DERIVED

Many questions begin by asking the candidate to explain a term of some sort and the term is often a physical quantity. You need to familiarise yourself with these physical quantities by referring to them frequently. Read the following list every so often and remind yourself of the interconnections and the definitions.

This appendix lists most of the different physical quantities included in A level physics syllabuses with their units. They have been listed so that, apart from the occasional exception, the definition of any physical quantity or its unit depends only on quantities listed above it. Also, the fundamental role of the base physical quantities is emphasised.

The scalar or vector nature of a physical quantity is listed alongside each entry. Pseudovectors (rotational vectors like torque) are not on any syllabuses; the labels have been included here to eliminate them from lists of scalars or vectors.

Length in metres	l	m	scalar
angle	θ	rad, °	pseudovector
distance	d	m	scalar
distance travelled	d	m	scalar
displacement	s, x	m	vector
amplitude	x_0, y_0	m	scalar
wavelength	λ	m	scalar
area	A	m^2	pseudovector
volume	V	m^3	scalar
Time in seconds	t	s	scalar
frequency	f	Hz	scalar
angular frequency, angular speed	ω	rad s^{-1}	scalar
speed	c	m s^{-1}	scalar
rate of change of speed		m s^{-2}	scalar
uniform, average, instantaneous velocities	u, v	m s^{-1}	vectors
acceleration	a	m s^{-2}	vector
acceleration of free fall	g	m s^{-2}	vector
gravitational field strength	g	N kg^{-1}	vector

Mass in kilograms	m	kg	scalar
density	ρ	kg m^{-3}	scalar
momentum	p, mv	kg m s^{-1}	vector
force	F	N	vector
impulse	Δp	N s	vector
moment of force	M	N m	pseudovector
torque	T	N m	pseudovector
pressure	p	Pa	scalar
stress	σ	Pa	scalar
strain	ε	no unit	scalar
Young's modulus	E	Pa	scalar
gravitational constant	G	N kg^{-2} m^2	scalar
work	W	J	scalar
energy	W, E	J	scalar
kinetic energy	E_k	J	scalar
potential energy	E_p	J	scalar
power	P	W	scalar
Temperature in kelvin	T	K	scalar
internal energy	U	J	scalar
heating	q	J	scalar
heat capacity	C	J K^{-1}	scalar
specific heat capacity	c	J kg^{-1} K^{-1}	scalar
specific latent heat (specific enthalpy)	L	J kg^{-1}	scalar
temperature gradient	$\Delta T/\Delta x$	K m^{-1}	scalar
thermal conductivity	λ	W m^{-1} K^{-1}	scalar
U value	U	W m^{-2} K^{-1}	scalar
Amount of substance	n	mol	scalar
molar mass	M	kg mol^{-1}	scalar
molar heat capacity	C_m	J mol^{-1} K^{-1}	scalar
molar gas constant	R	J mol^{-1} K^{-1}	scalar
Avogadro constant	N_A	mol^{-1}	scalar
Boltzmann constant	k	J K^{-1}	scalar
Current	I	A	scalar
charge	Q	C	scalar
electric field strength	E	N C^{-1}, V m^{-1}	vector

electric potential	V	V	scalar
permittivity of free space	ε_0	F m^{-1}	scalar
relative permittivity	ε_r	no unit	scalar
permeability of free space	μ_0	H m^{-1}	scalar
potential difference	V	V	scalar
electromotive force	E	V	scalar
capacitance	C	F	scalar
resistance	R	Ω	scalar
resistivity	ρ	Ω m	scalar
conductance	G	S	scalar
conductivity	σ	S m^{-1}	scalar
inductance	L	H	scalar
reactance	X	Ω	scalar
impedance	Z	Ω	scalar
time constant	τ	s	scalar
magnetic field strength	B	T	vector
magnetic flux	ϕ	Wb	scalar
magnetic flux linkages	$N\phi$	Wb turn	scalar
Curie temperature	T_c	K	scalar
Planck constant	h	J s	scalar
photon energy	hf	J	scalar
work function	ϕ	J	scalar
activity	A	Bq	scalar
decay constant	λ	s^{-1}	scalar
half-life	$T_{1/2}$	s	scalar
mass excess	Δm	kg, u	scalar
binding energy	E_b	J	scalar
binding energy per nucleon	E_b/A	J nucleon^{-1}	scalar
mass number	A		
atomic number, proton number	Z		
neutron number	N		

There are one or two tricky points. Some scalar quantities like current and e.m.f. become vectors in more advanced work. They are scalars at A level. The vector nature or otherwise of the pseudovectors should not be tested. Whether or not a physical constant such as the Boltzmann constant is a scalar depends on the strict definition of a scalar. They are certainly not vectors.

USEFUL FORMULAE AND EQUATIONS

The formulae you may need in an A level examination are set out here in two lists. The first list contains those formulae which you are meant to know and remember when you go into the examination. It also includes many of the word equations you will need. The second list contains those formulae which are provided in the formulae sheets. Both lists are as comprehensive as possible and may, therefore, include a few formulae not required or printed by your particular examination board. The first list is rooted in the 'Subject Core for Physics' and is much the same for all boards.

List A: You must know these

pressure = $\dfrac{\text{force}}{\text{area}}$ $\qquad\qquad p = F/A$

speed = $\dfrac{\text{distance travelled}}{\text{time taken}}$ $\qquad\qquad v = \Delta s/\Delta t$

acceleration = $\dfrac{\text{increase in velocity}}{\text{time taken}}$ $\qquad\qquad a = \Delta v/\Delta t$

momentum = mass × velocity $\qquad\qquad p = mv$

resultant force = mass × acceleration $\qquad\qquad F = ma$

weight = mass × gravitational field strength $\qquad\qquad W = mg$

inverse square laws of force: $\qquad\qquad F = G\dfrac{m_1 m_2}{r^2} \qquad F = \dfrac{Q_1 Q_2}{4\pi\varepsilon_0 d^2}$

work done = force × distance moved in the direction of the force $\qquad\qquad W = Fs \cos\theta$

power = $\dfrac{\text{energy transferred}}{\text{time taken}} = \dfrac{\text{work done}}{\text{time taken}}$ $\qquad\qquad P = \dfrac{\Delta E}{\Delta t} = \dfrac{\Delta W}{\Delta t}$

kinetic energy = ½ mass × velocity^2 $\qquad\qquad E_k = \tfrac{1}{2} mv^2$

change in potential energy =
  mass × gravitational field strength
  × change in height

$$E_p = mg\Delta h$$

centripetal force $= \dfrac{\text{mass} \times \text{speed}^2}{\text{radius}}$

$$F = mv^2/r$$

moment of force F about point O
  = $F \times$ perpendicular distance from F to O

$$M = F \times d$$

sum of clockwise moments about any point in a plane = sum of anticlockwise moments about that point

pressure × volume = number of moles
  × molar gas constant
  × absolute temperature

$$pV = nRT$$

$$\dfrac{P_1 V_1}{T_1} = \dfrac{P_2 V_2}{T_2}$$

wave speed = frequency × wavelength

$$c = f \times \lambda$$

current = charge / time interval

$$I = \Delta Q/\Delta t$$

potential difference = work done / charge

$$V = W/Q$$

electrical energy transfer =
  current × potential difference × time interval

$$\Delta W = IV\Delta t$$

electrical power = potential difference × current

$$P = VI = I^2 R$$

resistance $= \dfrac{\text{resistivity} \times \text{length}}{\text{cross-sectional area}}$

$$R = \dfrac{V}{I} = \dfrac{\rho l}{A}$$

capacitance $= \dfrac{\text{charge stored}}{\text{potential difference}}$

$$C = \dfrac{Q}{V}$$

$\dfrac{\text{input voltage}}{\text{output voltage}} = \dfrac{\text{primary turns}}{\text{secondary turns}}$

$$\dfrac{V_P}{V_S} = \dfrac{N_P}{N_S}$$

List B: Equations on formulae sheets (Check your own Board)

Mechanics

$$s = \dfrac{u + v}{2} t$$

$$\omega = \dfrac{\Delta\theta}{\Delta t} = \dfrac{v}{r}$$

$$T = 2\pi\sqrt{\left(\dfrac{m}{k}\right)}$$

$$v = u + at$$

$$T = \dfrac{1}{f} = \dfrac{2\pi}{\omega}$$

$$\rho = \dfrac{m}{V}$$

$$s = ut + \tfrac{1}{2} at^2$$

$$a = \omega^2 r = \frac{v^2}{r}$$

$$F = k\,\Delta x$$

$$v^2 = u^2 + 2as$$

$$x = x_0 \cos \omega t$$

$$\sigma = \frac{F}{A}$$

$$F = \frac{\Delta p}{\Delta t}$$

$$v = -\omega x_0 \sin \omega t$$

$$\varepsilon = \frac{\Delta l}{l}$$

$$F\Delta t = \Delta(mv)$$

$$a = -\omega^2 x$$

$$E = \frac{\sigma}{\varepsilon} = \frac{Fl}{A\Delta l}$$

$$P = F \times v$$

$$T = 2\pi\sqrt{\left(\frac{l}{g}\right)}$$

$$W = \tfrac{1}{2}\,F\Delta x = \tfrac{1}{2}\,k\Delta x^2$$

$$T = I(\omega_2 - \omega_1)/t$$

$$W = T\theta = \Delta(\tfrac{1}{2}\,I\omega^2)$$

Thermal physics

$$\theta/°C = T/K - 273.15$$

$$C_p - C_V = R$$

$$R = kN_A$$

$$\theta = \frac{X_\theta - X_0}{X_{100} - X_0} \times 100°C$$

$$\frac{\Delta Q}{\Delta t} = kA\frac{\Delta T}{\Delta x}$$

$$\Delta p = \rho g\Delta h$$

$$\Delta W = -p.\Delta V$$

$$\frac{\Delta Q}{\Delta t} = UA\Delta T$$

$$\eta = \frac{Q_1 - Q_2}{Q_1} = \frac{T_1 - T_2}{T_1}$$

$$\Delta U = mc\Delta T$$

$$pV = \tfrac{1}{3}\,Nm\langle c^2\rangle$$

$$\Delta U = L.\Delta m$$

$$T \propto \langle molecular\ k.e.\rangle$$

$$\Delta U = \Delta Q + \Delta W$$

$$\langle molecular\ k.e.\rangle = \tfrac{3}{2}kT$$

Waves, quantum phenomena and radioactivity

$$I = \frac{P}{4\pi r^2}$$

$$f = \frac{1}{2l}\sqrt{\left(\frac{T}{\mu}\right)}$$

$$A = \frac{dN}{dt} = -\lambda N$$

$$\lambda = \frac{xs}{D}$$

$$\Delta E = c^2.\Delta m$$

$$N = N_0 e^{-\lambda t}$$

$$n\lambda = d \sin \theta$$

$$E = hf$$

$$T_{1/2} = \frac{0.693}{\lambda}$$

$$\frac{\sin \theta_1}{\sin \theta_2} = \frac{\lambda_1}{\lambda_2} = \frac{c_1}{c_2} = \frac{n_2}{n_1} = {}_1n_2$$

$$hf = \phi + (\tfrac{1}{2}mv^2)_{max}$$

$$\sin \theta_c = \frac{c_2}{c_1} = \frac{1}{{}_1 n_2} \qquad \lambda = \frac{h}{p} = \frac{h}{mv}$$

$$n_1 = \frac{c}{c_1} \qquad hf = E_1 - E_2$$

Current electricity

$$I = nAQv \qquad W = \tfrac{1}{2}CV^2 \qquad I = I_0 \sin 2\pi ft$$

$$V = E - Ir \qquad Q = Q_0 e^{-t/CR} \qquad I_{r.m.s.} = I_0/\sqrt{2}$$

$$R = R_1 + R_2 + R_3 \qquad Q = Q_0(1 - e^{-t/CR}) \qquad V_{r.m.s.} = V_0/\sqrt{2}$$

$$\frac{1}{R} = \frac{1}{R_1} + \frac{1}{R_2} + \frac{1}{R_3} \qquad \tau = CR \qquad f_{resonance} = \frac{1}{2\pi\sqrt{(LC)}}$$

$$\frac{1}{C} = \frac{1}{C_1} + \frac{1}{C_2} + \frac{1}{C_3} \qquad X_C = 1/\omega C = 1/2\pi fC \qquad <P> = I_{r.m.s.} V_{r.m.s.} = I_0 V_0/2$$

$$C = C_1 + C_2 + C_3 \qquad X_L = \omega L = 2\pi fL$$

Fields

$$E = F/Q \qquad F = BIl \qquad E = -N\frac{d\phi}{dt}$$

$$E = V/d \text{ (uniform)} \qquad F = BQv \qquad E = Blv$$

$$E = Q/4\pi\varepsilon_0 r^2 \text{ (radial)} \qquad B = \mu_0 I/2\pi r \qquad E = \omega BAN \sin 2\pi ft$$

$$V = Q/4\pi\varepsilon_0 r \text{ (radial)} \qquad B = \mu_0 NI/l = \mu_0 nI$$

$$e.\Delta V = \Delta(\tfrac{1}{2}m_e v^2) \qquad B = \mu_0 NI/2r \qquad g = F/m$$

$$C = \varepsilon_0 \varepsilon_r A/d \qquad F = \frac{\mu_0 I_1 I_2}{2\pi r} \qquad g = Gm/r^2 \text{ (radial)}$$

$$\phi = BA \qquad V = -Gm/r \text{ (radial)}$$

PHYSICAL CONSTANTS AND USEFUL DATA

Physical quantity	Symbol	Value	Unit
speed of light in vacuum (exact value − defined)	c	$2.997\ 924\ 58 \times 10^8$	$m\ s^{-1}$
permeability of vacuum (exact value − defined)	μ_0	$4\pi \times 10^{-7}$	$H\ m^{-1}$
permittivity of vacuum (defined $= 1/\mu_0 c^2$)	ε_0	$8.854\ 187\ 817 \times 10^{-12}$	$F\ m^{-1}$
gravitational constant	G	6.673×10^{-11}	$m^3\ kg^{-1}\ s^{-2}$
molar gas constant	R	8.315	$J\ K^{-1}\ mol^{-1}$
Avogadro constant	L or N_A	6.022×10^{23}	mol^{-1}
Boltzmann constant ($= R/L$)	k	1.381×10^{-23}	$J\ K^{-1}$
elementary charge	e	1.602×10^{-19}	C
Faraday constant ($= eL$)	F	9.649×10^4	$C\ mol^{-1}$
Stefan–Boltzmann constant	σ	5.671×10^{-8}	$W\ m^{-2}\ K^{-4}$
Wien's law constant	c_w	2.898×10^{-3}	$m\ K$
unified atomic mass constant	u or m_u	1.661×10^{-27}	kg
electron rest mass	m_e	9.109×10^{-31}	kg
neutron rest mass	m_n	1.675×10^{-27}	kg
proton rest mass	m_p	1.673×10^{-27}	kg
specific charge of electron	e/m_e	1.759×10^{11}	$C\ kg^{-1}$
Planck constant	h	6.626×10^{-34}	$J\ s$
Rydberg constant	R_∞	1.097×10^7	m

Physical quantity	Symbol	Value	Unit
acceleration of free fall near Earth's surface	g	9.81	$m\ s^{-2}$
gravitational field strength near Earth's surface	g	9.81	$N\ kg^{-1}$
standard pressure		100 000	Pa or 1 bar
standard atmosphere		$1.013\ 25 \times 10^5$	Pa
standard temperature		273.15	K or 0°C
ice point		273.15	K or 0°C
triple point of water		273.16	K
density of dry air at s.t.p.		1.293	$kg\ m^{-3}$
density of water at s.t.p.		1.000×10^3	$kg\ m^{-3}$
molar mass of air		0.028 98	$kg\ mol^{-1}$
molar volume at s.t.p. ($= RT/p$)		22.414×10^{-3}	m^3
molar volume at standard atmospheric pressure and 298 K		24.467×10^{-3}	m^3
radius of Sun		6.96×10^8	m
mass of Sun		1.99×10^{30}	kg
power radiated from Sun		3.85×10^{26}	W
temperature of Sun's surface		5770	K
distance of Sun from Earth		1.496×10^{11}	m
radius of Earth at equator		6.378×10^6	m
radius of Earth at poles		6.357×10^6	m
mass of the Earth		5.976×10^{24}	kg
standard gravitational acceleration at surface	g	9.807	$m\ s^{-2}$
radius of the Moon		1.739×10^6	m
mass of the Moon		7.36×10^{22}	kg
distance of Moon from Earth		3.844×10^8	m

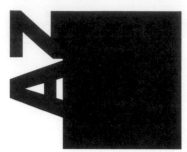

EXAMINERS' TERMS

We are all familiar with the advice to 'read the question'. And if reading the question were enough to decide what we should write, then all would be well and good. 'Explain' would mean explain and 'define' would mean define. But reading the question is never enough. Examiners do not respect particular meanings of these words as if they were some kind of code. You must interpret the words in context.

Explain sometimes means explain, it sometimes means describe and it sometimes means define. Similar points can be made about other introductory terms. This is not as confusing as it sounds. If the question is read carefully along with the mark allocation, the meaning of the question will be clear.

Now read the following notes about frequently-used examination terms.

Define

Define always means write out a definition. A word equation is usually sufficient and often easier to write out accurately. If you use an algebraic definition remember to define the symbols in your equation.

Explain the meaning of ... Explain what is meant by ...

A definition is sometimes enough. If the mark allocation suggests that a simple definition isn't enough, then give an example. Maybe draw a diagram. The term 'gradient' for example, is better explained with a sketch graph than with a few fine words. Terms like 'equilibrium' are best explained by giving an example.

Describe ...

Have a look at a few questions. You won't be asked to describe a helical spring or an electric field or a set of expensive dinner plates. You will be asked to describe a process like the energy transformations in a simple pendulum or an experiment to illustrate something. Say how things are at the start, in the middle and at the end. Concentrate on getting the sequence right. The mark allocation is the best clue to how many steps you should include. Draw a labelled diagram whenever possible.

Explain why ...

The question will indicate the happening that needs explanation. The explanation will always involve a principle of some sort. Identify and state the relevant principle, e.g. impulse equals change of momentum or total charge stored is constant. Apply the principle to the initial state of the situation and show how it leads to the outcome.

Explain what happens when ...

State clearly what happens and then 'Explain why ...'

Calculate ... or Find ...

These words ask you to work out a numerical problem. State clearly the principle or the basic equation. Re-arrange the equation. Put in the data in SI units, not their multiples or sub-multiples. Calculate the result. A tidy calculation can score most of the marks even when the answer is wrong. A jumble of figures and a wrong answer scores nothing.

Use as many significant figures as you like but reduce the number to match the initial data when you state the answer. Never forget the unit.

Estimate ...

Estimate usually means that you supply some typical data yourself and calculate a typical value. Sometimes all the data is in the question but you need to make an approximating assumption – assume, for example, that steam behaves like a perfect gas or that a frictional force is zero.

State ... or List ...

These words should be taken literally.

Sketch ...

Whether it is a graph that needs sketching or apparatus for an experiment, the marks will always be given to the labels.

When 'sketching a graph', label the axes, get the shape of the curve roughly correct and put in any magnitudes you can, such as the value of an intercept or the meaning of a gradient. If the axes cross at the origin, then mark the origin 0,0. Don't let a decay graph touch the axis.

When 'sketching apparatus', draw the two or three important parts first in roughly the right places and then join them up and label them. Avoid using rulers: the lines are rarely in the right places, they look wrong, the rulers take up valuable time and the marks are given for correct labels not for accurate drawing.

REVISION LIST

This appendix should help you to make best use of the book. It is not an exhaustive list of topics in your syllabus. It is a list of key starting points. Look up any one of them in the text and the cross-referencing will deepen and extend your grasp of that syllabus area.

1 Physical quantities

base quantity

derived quantity

gradient

graph

homogeneity

limiting value

rate of change

scalar quantity

uncertainty

vector quantity

2 Force and motion

acceleration of free fall

displacement–time graph

equations of motion

free-body force diagram

kinds of force

Newton's laws of motion

power

velocity–time graph

weightlessness

work

3 Circular motion

angular speed

centripetal force

circular motion

couple

forces in equilibrium

frequency

gravitational constant G

Newton's law of gravitation

principle of moments

satellite

4 Simple harmonic motion

amplitude

damped harmonic motion

forced oscillations

frequency

helical spring

period

resonance

simple harmonic motion

simple pendulum

5 Matter

Archimedes' principle

Boyle's law

elastic limit

energy–separation curve

force–extension curves

ideal gas equation

kinetic theory of gases

polymeric solids

pressure

states of matter

6 Thermal physics

finite energy sources

heat capacity

heat engine concept

heating and working

internal energy

specific latent heat

thermal conduction

thermal equilibrium

thermal fission reactor

thermodynamics (first law of)

7 Current electricity

ampere

coulomb

current–voltage characteristics

electrical power

e.m.f.

internal resistance

Ohm's law

resistivity

resistors in parallel

resistors in series

8 Sound waves

frequency

longitudinal waves

loudness

nodes

pitch

progressive waves

sound waves

speed of sound

stationary wave

superposition

9 Light and electromagnetic waves

diffraction

electromagnetic spectrum

interference

light as a transverse wave motion

light ray

polarisation phenomena

reflection of light

refraction of light

total internal reflection

two slit experiment

10 Electric and magnetic fields

capacitance

capacitors in AC circuits

Coulomb's law

electric field strength

energy stored in a capacitor

force on moving charge

magnetic effect of current

magnetic flux density

parallel plate capacitor

time constant

11 Electromagnetic induction and alternating current

alternating current and voltage

capacitors in AC circuits

electromagnetic induction (laws of)

flux linkage

inductive reactance

magnetic flux

mean power in resistive load

rectification: half-wave and full-wave

r.m.s. (root mean square)

transformer

12 Radiation and matter

energy level

excitation

hydrogen atom

ionisation

nuclear atom

photoelectric effect

photon

threshold frequency

wave–particle duality

work function energy

13 Radioactivity

activity

alpha radiation

background radiation

becquerel

beta radiation

decay constant

exponential law of radioactive decay

gamma radiation

half-life

radioactive decay

14 The nuclear atom

alpha particle scattering experiment

atomic number

binding energy per nucleon

fission

isotope

N–Z curve for stable nuclei

mass number

nuclear reactions

nucleus

unified atomic mass constant

REVISION LIST

15 Materials

composite materials
crystalline solids
dislocations
elasticity
glasses
hardness
helical spring

metals
polymeric solids
stiffness
strength
stress–strain curves
toughness
Young modulus

16 Medical physics

ear
eye
MRI
radiation dosage

radioisotopes
ultrasound
X-rays

17 Particle physics

antimatter
baryons
exchange particles
Feynman diagrams
four fundamental interactions
hadrons

leptons
mesons
particle accelerators
particle interactions
particle physics
quarks

18 Astronomy

astrophysics
CCDs
cepheid variable
end states of stars
galaxies

Hertzsprung–Russell diagram
parallax (stellar)
star
telescopes

19 Telecommunications

analogue signals
analogue transmission systems
digital signals
digital transmission systems
frequency division multiplexing
frequency modulation

noise
Nyquist's theorem
pulse code modulation
sampling
time division multiplexing

SYNOPTIC QUESTIONS IN A LEVEL PHYSICS

There is a general requirement that A level syllabuses for courses starting in September 2000 should include a synoptic element in the examinations. This requirement will be interpreted a little differently from subject to subject.

Synoptic, as far as physics is concerned, means that the questions in one part of the examination will call on knowledge from more than one area of the syllabus without destroying the unity of the question. Before you decide that this is all too difficult, remember that no one – government, universities, schools, parents, candidates – will thank an examiner who sets questions that candidates can't answer. The substance of the following notes will be clear to you in the later stages of your course. If you are just starting the course, you may well find them depressing. Don't worry.

How will the demand for synoptic questions be managed? We can cover most of the possibilities under three headings: mathematical questions, topic questions and comprehension questions.

Mathematical questions

The examiner exploits two situations which involve similar mathematical equations; not difficult, just similar. The following examples are well known:

1   Inverse square laws. There are two inverse square laws of force on all syllabuses, gravitational and electrical. Similarities include the vector nature of the force, the scalar nature of the potential (unlikely to be examined) and a uniform field on a local scale. In both cases the force acts along a radius; there is no component along a tangent. Differences largely centre on electric fields being either attractive or repulsive and gravity being the important force at the astronomical level because matter in sufficient bulk in electrically neutral. Another important difference is that charge can be moved around and stored on conductors, thereby creating the possibility of manipulating the shape of the field.

There is a second set of inverse square laws, the laws connected with particle or radiation flow. Any electromagnetic radiation from a point source weakens inversely with the square of the distance. It happens too with concentrated alpha and beta radiation sources in vacuum and with the fabled butter gun.

2   Exponential decay. Exponential decay occurs whenever the rate at which some quantity diminishes is in proportion to its present value. This point is stressed in the diagram for *exponential decay*. Three situations likely to be tested are radioactive decay, capacitor

discharge through a resistor and damped harmonic motion. Similarities involve the causes – activity proportional to undecayed mass, discharge current proportional to charge stored and work done against friction per cycle proportional to the square of amplitude. Be ready to compare activity and half-life on the one hand with discharge current and time constant on the other. A graph of ln(activity) against time or ln(discharge current) against time enables half-life or time constant to be evaluated.

A question could pick up on the contrast between the exponential decay with distance brought about by the absorption of, say, γ-rays in aluminium and the exponential decay with time brought about by radioactive decay. Both ideas are within your syllabus.

3   Stored energy. Compare $\frac{1}{2}kx^2$ for a helical spring with $\frac{1}{2}CV^2$ for a capacitor. Both laws derive from a linear property of the hardware, $F = kx$ and $Q = CV$. Both are important because the energy can be recovered without transforming into the disordered form (what used to be called heat). In terms of the first law of thermodynamics, ΔW is transferred to the system and ΔU increases. It is the other way round when the energy is recovered. ΔQ is zero at all times. This last point can turn up with non-ideal systems; ΔQ is not zero and not all of the input energy is recoverable.

Synoptic topics

The examiner chooses a topic which impinges on the syllabus in more than one area. The only way to prepare for this style of question is to watch for possibilities while you are working through the course. They mostly turn up towards the end. Common examples are found with important instruments. The mass spectrometer involves a velocity selector and, like the cyclotron, a centripetal force which derives from the motion of a charged particle in a magnetic field. All radiation detectors rely on absorption processes. A question on the nuclear fission reactor can drift into thermal efficiency, energy conversion at the nuclear level (Δmc^2) and energy conversion processes in the turbines, generators, transformers and the transmission system. The idea of efficiency is relevant to anything involving energy conversion processes, from X-ray tubes to electric motors. It is always the sources of *inefficiency* that matter.

Comprehension passages

These questions do no more than test whether or not you have made an honest attempt to cover the syllabus. A familiar topic is tested in a quaint situation, like wet hands under a hand drier switching on the drier by changing the capacitance of a capacitor inside the box. Why do you suppose the hands have to be underneath and not to one side? You can't prepare for the particular question you are going to get but you can be prepared. Know exactly what you are going to do.

Read each part of the question carefully and identify the part of the passage it relates to – is the answer in the passage or do you have to involve part of what you have learned across the syllabus? When you have finished writing out your answers you will know much more about the passage. Perhaps your earlier answers will need to be changed a little. Leave yourself a bit of space to do this.

PREPARING FOR THE EXAMINATION

This appendix begins with some ideas for planning your revision, it continues with some notes on how to answer examination questions or work through calculations and ends with some detailed discussion about describing experiments.

Planning your revision

Students always have their own preferences and their own work patterns. However you manage your own affairs, don't do less than is recommended here.

Your revision period starts about two months before the examination. This is still true if you are taking just one module; the brain needs time to mull over the important ideas, to become familiar with all the details and to be fluent with its responses. One module will not need so much time per day but it will need the same stretch of time.

Begin your revision by dividing the syllabus into bite-sized pieces. About 30 of these will cover the complete syllabus. Your list will look something like this:

- physical quantities and their units

- straight line motion

- force, motion, energy and power

- motion in a circle

- simple harmonic motion

  .
  .
  .

- photoelectric effect

- radioactivity

  .
  .
  .

Get some small white cards and, as you revise each area of the syllabus, list what you need to know about that topic on its own card.

The simple harmonic motion card should include these reminders of definitions, formulae, graphs and experiments:

- s.h.m. – definition (mag and dir)

- definitions – amplitude, frequency, period, restoring force, spring constant
- formula sheet – x, a, T, max speed and acc'n
- know formulae for v, E_k
- graphs – x, v, a, F, E_k against time, a, F and E_k against displacement
- trace conservation of energy through energy transformations
- energy variation with displacement for simple pendulum and helical spring
- experiments: simple pendulum, helical spring

and the photoelectric effect card will have a similar set of reminders:

- photoelectric effect – definition
- definitions – work function, stopping potential, threshold frequency
- photoelectron, photon, photon energy hf.
- Einstein's photoelectric equation
- Experiment – vacuum, incident photons, collector grid, electrometer or nA,
- voltmeter
- Graphs – stopping potential against frequency, gradient = h/e, ...– photoelectric current against grid voltage, vary incident intensity ... – photoelectric current against incident intensity, grid voltage fixed.

Write out these cards as you work through the syllabus section by section. Go back to the syllabus and make sure nothing is left out. Read the cards frequently and keep your grip on each syllabus section you have worked through. Always learn whole sections, and don't leave things out.

Answering questions

Questions in modular examinations are broken down into short sections. All the candidate has to do is write out the answers carefully. But this is easier said than done and it is worthwhile asking why so many candidates end up with lower marks than they think they deserve.

Consider the following imaginary question:

- Explain why a metal block gets hot when it is rubbed on a concrete floor. [2 marks]

'Look at the price before you listen to the salesman'. Two marks means you have to include two distinct points in your answer. The candidate who writes 'Because the energy transferred by working against friction is transformed into heat' gets one mark. The second mark will go for something along the lines of 'This heat flows into the metal block because the metal is a much better thermal conductor than the concrete'. Half of candidates or more will lose that second mark. Don't omit an essential step because it seems too obvious.

Many questions involve a contrast. Consider

- Why does A warm up faster than B? [2 marks]

Your answer must include a statement about A, a statement about B and a reason for the difference. 'Because A is inside the furnace' gets one mark. 'Because A is inside the furnace where it is hot but B stays outside the furnace where it is cool' gets two marks.

There would be many candidates who could answer these last two questions, but who would earn only two out of the four marks. An average candidate will repeatedly lose marks by giving only half of any answer, often thinking that the rest of the answer is too obvious to bother about. Don't fall into this trap. You can drop by as much as two or three grades through not paying attention to the details in your answers. You may not like this way of doing things but you have no choice if you want to earn a high grade.

If, in answer to a question such as 'define resistivity' or 'state the law of universal gravitation' you write down an equation without stating what the symbols mean, you score nothing. List the symbols and their meanings immediately below equations. Use word equation definitions wherever possible: they avoid many of these problems.

Often, you are asked to sketch a graph. There will be one mark for labelling the axes with variables and their units and, possibly, for marking the scale values along the axes. There will be another mark for having the shape of the curve roughly correct (don't let a decay curve cut the time axis). There will also be one mark for giving the numerical value of an intercept if the information is available and one mark for stating what the gradient equals or what an identified area equals. These are the items you should try to include on your sketch.

There is an increasing habit of including a graph on a question paper and asking candidates to use it to estimate the value of some quantity. The word 'estimate' is used to indicate that answers will vary somewhat from one candidate to another. These questions mostly require either the evaluation of a gradient or the evaluation of an area. To evaluate a gradient, draw the tangent carefully and at the right point along the curve. Draw as large a triangle as possible and measure the rise and the step according to the scales on the axes parallel to the lines. Don't forget the unit; it is the ratio of the y-axis unit to the x-axis unit. Estimating an area is often a little troublesome. If the easiest way of finding the relevant area is to count squares, then count squares. But don't forget to use the axes scales to evaluate the area of one square. The unit for the area is the product of the units on the two axes.

When you are asked to draw a diagram, the marks will be given for correctly labelled parts of the diagram. A part which is not labelled, however obvious its meaning, will not get a mark.

Calculations

Many calculations are short and you will get all the marks if you get the right answer and remember to include the unit. There is no problem with that, but you need to guard against losing marks unnecessarily when you don't get the right answer. This is why your method needs to be clear without being over-detailed.

Marking schemes vary occasionally, but usually there is one mark for a correct statement of principle, one mark for manipulating the equation, another for substituting the data correctly and one for the right answer with its unit. Where there are only three marks, it is the middle two marks that are combined.

Candidates who don't know the principle on which a calculation is based can't start on the calculation. It is all too late for these candidates. More worrying is the failure of other candidates to put the data correctly into an equation. Equations are correct provided that the data is in standard SI units. Current must be in amps, not milliamps. Mass must be in kilograms, not grams. Be very careful with areas. A radius of 2.2 cm gives an area of $\pi.(2.2 \times 10^{-2})^2$ m^2 and 8.1 mm gives an area of $\pi.(8.1 \times 10^{-3})^2$ m^2. These last two calculations often turn up incorrectly. Candidates frequently replace 1.0 mm^2 with 1.0×10^{-3} m^2 instead of 1.0×10^{-6} m^2.

Describing experiments

It is rare to find an examination paper which does not include one question largely given over to a description of an experiment. A typical question gives you ten minutes or so to earn about eight marks. The way that the marks are distributed will, of course, vary from one occasion to another but the broad pattern is always the same.

Follow these rules:

Apparatus

Put as much information as you can into a fully labelled diagram. Labels need to be brief but informative, e.g. 'timer runs while this light beam is broken', 'fixed mass of gas', 'grid at low negative potential to slow down and repel electrons'. Identify precisely any length, area or volume that is going to be measured.

Measurements

List each measurement and name the instruments used, e.g. 'measure the length of the pendulum with a metre rule', 'add the barometer reading to the pressure gauge reading to find the gas pressure', 'measure the terminal voltage with the digital voltmeter and, at the same time, record the current in the ammeter'. Whenever possible, state that a measurement is repeated so that an average value can be worked out and used.

Conclusion

Frequently, an experiment concludes by requiring the candidate to draw a graph and to check whether or not it is straight, whether or not it goes through the origin or to find a gradient. Always sketch the graph you expect to get, label the axes and identify the feature of interest. Don't bother with rulers: straight lines waste time and they don't matter. Labels do.

The following list includes most of the experiments that examiners ask for. They are in no particular order. You need to check whether or not your Board excludes any of these experiments.

- Determination of centre of mass
- Hooke's law and the Young modulus
- Measurement of spring constant
- Use of a linear air track to check equations of motion, proportionality of force and acceleration, conservation of momentum in collisions, loss of mechanical energy in non-elastic collisions
- Check current–voltage curves
- Measurement of time constant
- Use of simple pendulum to measure g

- Measurement of **g** by free fall
- Measurement of the speed of sound in free air
- Measurement of half-life or attenuation coefficient
- Checking the inverse square law for gamma radiation
- Observation of Brownian movement
- Measurement of capacitance and dielectric constant
- Demonstration of the direction of an induced e.m.f.
- Verification of each of the three gas laws
- Use of a diffraction grating to measure wavelength
- Measurement of specific and latent heats (enthalpies)
- Use of a thermocouple
- Determination of internal resistance or of resistivity
- Measurement of charge stored on a capacitor; use of the coulombmeter
- Measurement of time constant
- Use of the Hall probe to check magnetic field strength formulae
- Verification of Einstein's photoelectric equation
- Determination of work function energy
- Rutherford's alpha particle scattering experiment
- Millikan's measurement of electronic charge
- Measurement of the gravitational constant
- Demonstrating the wave nature of the electron

This may all seem too much work for one question. But it isn't one question. If you take four or five theory papers, then you may be tested on as many as four or five of the experiments. Also, they spread right across a syllabus and come in at every crucial point. Learning these experiments will give you a general command of your syllabus and knowing about these experiments will help you with other parts of questions. It would probably be worth your while to summarise the details for each experiment on its own white card.

WORKING FOR AN A GRADE

A-grade candidates in physics do the ordinary things well and then go an extra distance. The ordinary things are a fluency with the main physical laws and all the basic definitions, the ability to work through the calculations quickly and accurately and the habit of writing out simple explanations clearly, correctly and briefly. These things get the high marks. The extra distance is about developing a stronger understanding of the syllabus as a whole by drawing together and linking up apparently different ideas from separate parts of the syllabus. This appendix identifies and highlights some of these links.

A The conservation laws in mechanical systems

Multiply both side of the equation $v - u = at$ by mass m and we get

$$mv - mu = (ma).t = F.t = \text{impulse}$$

That impulse equals change of momentum is a simple development of an equation of motion which is no more than a definition of uniform acceleration. Newton's third law of motion tells us that impulses, like forces, come in equal and opposite pairs. It follows that the total momentum of an isolated system is constant in the absence of external forces. But real systems are never isolated. Examiners expect you to identify external forces such as friction, which cause energy and momentum to leak from an apparently isolated system into the surroundings.

Multiply both sides of the equation $v^2 - u^2 = 2as$ by mass m and we get

$$\tfrac{1}{2}mv^2 - \tfrac{1}{2}mu^2 = (ma).s = F.s = \text{work}$$

The idea that work equals increase in kinetic energy comes from a second equation of motion.

Any increase in linear momentum in one direction is matched by an equal increase in the linear momentum of a second body in the opposite direction. There must be two bodies because of Newton's third law. Both bodies gain momentum, but in opposite directions. The energy exchange which occurs at the same time is not symmetrical. There is a transfer of energy: a loss and a gain. One body works, the other is worked upon. A-grade candidates must practise the kind of joined-up thinking shown in this sweep from the equations of motion to the general laws of physics.

Sometimes the absence of an energy exchange is important. Since gravitational forces do no work on planets that move in circular orbits, the gravitational force must be perpendicular to the orbit. If we add to the planet's motion a simple harmonic oscillation along the radius with a period that is identical to the period of the circular motion, the orbit becomes elliptical. The gravitational attraction is no longer perpendicular to the orbit; there is a continuous exchange of energy between the planet and the gravitational field and back again. The speed of the planet increases and decreases alternately. A-grade candidates will tie this broad understanding of planetary motion to their view of atomic structure and be aware of two modifications, an electric field instead of a gravitational field and the quantisation of angular momentum. A-grade candidates will notice too that the difference of scale is in proportion to the ratio of the electrostatic field constant to the gravitational field constant.

B Energy and the first law of thermodynamics.

The energy concept runs right through physics. The definition of energy is contained implicitly within the first law of thermodynamics which states that the increase ΔU in the internal energy of a system is the sum of two components, an ordered component ΔW and a disordered component ΔQ. The ordered component usually comes from mechanical or electrical work; the disordered component is usually a thermal energy transfer. There is an asymmetry between ordered energy and disordered energy. Ordered energy can be transformed completely into disordered energy, but only a fraction of the disordered energy can be transformed back again. This forms the basis for the second law of thermodynamics; it explain why a level of inefficiency is built into even perfect heat engines.

C Energy and electrical circuits

One of the principal laws of electrical circuit behaviour, Kirchhoff's second law, is no more than the law of energy conservation applied to current loops. Electromagnetic induction is a little more complicated. Electromagnetic induction has to be an experimental fact; it cannot be shown to follow from the earlier laws of electromagnetism. Once experiment tells us that electromagnetic induction happens, then insisting that energy is conserved in the process leads ineluctably to the two laws of electromagnetic induction – one for the direction of the induced e.m.f. and the other for its magnitude.

The ideas of e.m.f. and potential difference are closely related to the energy concept. The e.m.f. of a power supply equals the stored energy in joules transformed into electrical energy per coulomb of charge exiting from the positive terminal. Potential difference is the electrical energy in joules transformed into disordered energy in a resistor per coulomb of charge flowing through. Notice how, apart from units related to the coulomb, the electrical units are all drawn from mechanics. The link between these two areas of study is the first law of thermodynamics.

D Energy, radioactivity and nuclear physics

Estimating nuclear radius is yet another application of the law of energy conservation. An alpha particle making a direct approach to the nucleus of a gold atom comes to rest when the rise in electrical potential energy equals the initial kinetic energy of the alpha particle. The one unknown, the minimum distance between the centres of the alpha particle and the gold nucleus, can be calculated. This gives an accurate estimate of the diameter of the gold nucleus. (Alpha particle energy can be measured directly by measuring the temperature rise of an irradiated metal container.)

Another idea which runs through the whole syllabus is the abstract nature of the mathematics used to describe the laws of nature. Equations of motion are precise, experiment is approximate. And so it is with radioactivity. There is the real and the experimental on the one hand and the mathematical and the abstract on the other. It is worth a closer look. Consider the equations

$$dN/dt = -\lambda N \quad \text{and} \quad N = N_0 \exp[-\lambda t]$$

Even though t is the only parameter in these equations that can be measured directly, the smooth decay curve described by the second equation is seen everywhere.

The two quantities that are measured in the ordinary course of events are count rate (which is proportional to dN/dt) and half-life. Half-life is the link between laboratory work and the mathematical and exact picture presented by the two equations above. It follows that great care is needed to get the definition of half-life just right. Several independent measurements of the same half-life will have values scattered randomly about the mathematical or the average value. This is why the word 'average' must be included in the definition of half-life. It is this average value which is used to calculate the decay constant λ and to draw the smooth decay curves which are so familiar.

E Energy and wave motion

Sound waves, water waves or electromagnetic waves are all examples of energy flow. Few aspects of wave motion can be discussed without reference to energy conservation. We go back to our work on simple harmonic motion to show that intensity is proportional to amplitude squared. Conservation of energy leads straight to the inverse square law for intensity. Similar reasoning accounts for the energy distribution in an interference pattern.

Since sound waves are mechanical waves, ideas about resonance and energy transfer are fundamental to our understanding of sources of sound generally and musical instruments in particular.

Einstein's photoelectric equation is no more than the law of conservation of energy applied to photons and electrons during a photoelectric event. Photon energy hf accounts for the chemical damage brought about by bright sunlight and ultra-violet light, and for the physical damage associated with much shorter wavelengths.

F Exponential decay – a mathematical form

Exponential decay is relevant to damped harmonic motion, the cooling of a hot body in surroundings at a lower but constant temperature, capacitor discharge and radioactive decay. In each of these situations, the equation driving the change has the same simple form. The matter is discussed in more detail in *exponential decay* and in Appendix 6.

G Fields

A-grade candidates should spend some quality time thinking about gravitational, electrical and magnetic fields. The gravitational field strength and the gravitational potential near an isolated point mass follow the same pattern as the electric field strength and electric potential near an isolated negative charge. With an isolated positive charge the direction of the electric field strength is reversed and potential decreases with increasing distance from the charge. The magnetic field near a long straight wire is described by circular lines of force, lines which neither begin nor end. There are strong similarities between the three kinds of field. There are also strong differences – especially in the ways in which the fields interact with matter.

A-grade candidates should deepen their understanding of the fundamentals of the subject matter within each syllabus section in turn and across the whole syllabus in the ways suggested above. If they can couple this with some care in the use of language then A-grade success is assured.

WORKING FOR AN A GRADE